Springer Series
in Physical Environment

7

J. Mangelsdorf   K. Scheurmann   F.-H. Weiß

River Morphology

A Guide for Geoscientists and Engineers

With 117 Figures

Springer-Verlag Berlin Heidelberg New York
London Paris Tokyo HongKong

Dr. JOACHIM MANGELSDORF
Bayerisches Landesamt für Wasserwirtschaft
Lazarettstraße 67
8000 München 19, FRG

Professor Dr.-Ing. KARL SCHEURMANN
Brüder-Grimm-Straße 14
8300 Landshut, FRG

Dipl.-Ing. FRITZ-HEINZ WEISS
Bayerisches Landesamt für Wasserwirtschaft
Lazarettstraße 67
8000 München 19, FRG

Translated from German by:

BARBARA REIMER
Bernhard-May-Straße 43
6200 Wiesbaden-Biebrich, FRG

Title of the original German edition:
Joachim Mangelsdorf / Karl Scheurmann
Flußmorphologie — Ein Leitfaden für Naturwissenschaftler und Ingenieure
© R. Oldenbourg-Verlag GmbH, München 1980

ISBN-13: 978-3-642-83779-1   e-ISBN-13: 978-3-642-83777-7
DOI: 10.1007/978-3-642-83777-7

Library of Congress Cataloging-in-Publication Data. Mangelsdorf, Joachim. [Flußmorphologie. English] River morphology: a guide for geoscientists and engineers / J. Mangelsdorf, K. Scheurmann, F. Weiss. p. cm. — (Springer series in physical environment; 7) Translation of: Flussmorphologie. Includes bibliographical references. ISBN-13: 978-3-642-83779-1 (U.S.:alk.paper) 1. River channels. I. Scheurmann, Karl. II. Weiss, F. (Fritz), 1935– . III. Title. IV. Series. GB561.M3613 1989 551.4′42 — dc20   89-26090

© Springer-Verlag Berlin Heidelberg 1990
Softcover reprint of the hardcover 1st edition   1990

Typesetting: K+V Fotosatz GmbH, Beerfelden

2132/3145-543210 — Printed on acid-free paper

Preface

River morphology is a multi-faceted branch of the geosciences and may be considered as a special topic of geomorphology as well as of fluvial sciences in general. It is thus closely connected with both hydrology and geology.

This book is essentially addressed to the specialist involved in hydraulic engineering, but also to the geographer, and to scientists in related fields, who are seeking information on the frequently very intricate processes of river bed formation.

The contents represent more a guide than a concise textbook. In addition to being an introduction, the book also serves another purpose: All natural phenomena form an intricate network of causal connections and are only as such accessible to scientific research. However, every science is subject to the danger of seeing only those aspects of reality which have been delineated by a specific method of research and which are the object of a specific research project. The observation of natural phenomena is at present frequently no longer object-oriented and direct, because the field is left rather to mathematical abstraction, which is thought to be the sole key to enlightenment.

Mathematical-statistical methods have gained a strong foothold in the geosciences, and, hydrology, for example, is today essentially controlled by them. Their usefulness can and should not be underrated − this book will frequently discuss indispensable mathematical approaches and calculations. Nevertheless, the direct observation of nature, and especially of the changes in the morphology of rivers and of their formation, is in danger of losing its importance.

In the geosciences this trend might well restrict a comprehensive view of the entire field. Geological and morphological processes usually proceed at extraordinarily slow speed and even river changes, which are short-lived compared to other geological processes, are scarcely perceptible to the human eye. Consequently, careful observation is an important corollary to the necessary measurements and their statistical evaluation.

The bibliography given with this book is an extensive one and aims at as broad as possible an overview of the present status of knowledge. The wealth of hydrological and geological/geomorphological literature − as in most other sciences − has become

overwhelming, but on closer perusal, the same problematic subjects appear repeatedly, resulting in a fairly even coverage. The
number of detailed descriptions of rivers and river basins was kept
low, to avoid overrepresentation.

The manuscript of the German-language issue was finalized in
February 1979. Before translating this first edition, the book was
entirely revised, completed, and extended by Chapter 8, River
Morphology. The authors will be grateful to have their attention
drawn to errors, omission of references, or gaps in themes which
one or the other reader might have wished to see treated in more
detail. Suggestions for improvement are greatly appreciated by the
authors.

The graphical lay-out was prepared by Mrs. B. Reiserer, assisted
by Mrs. G. Saurwein. Mrs. B. Kaufmann was of great assistance
in gathering the relevant literature from widely scattered sources.
The help of these ladies is gratefully acknowledged.

Thanks are also due to the publishers for accepting this text in
a widely distributed series of monographs. The authors are
grateful for the quality of presentation, and for the always fruitful
and pleasant cooperation.

Munich, Autumn 1989   J. MANGELSDORF
                                                    K. SCHEURMANN
                                                    F.-H. WEISS

Contents

List of Abbreviations and Units

A_E drainage area (km^2)

A_N area of precipitation (km^2)

$\bar{a}_c$ vector of Coriolis force ($m\,s^{-2}$)

a_c Coriolis force ($m\,s^{-2}$)

a_z centrifugal force ($m\,s^{-2}$)

b width of water surface (m)

b_G width of meander belt (m, km)

b_M meander amplitude (m, km)

b_S bottom width (m)

C_S concentration of suspended load ($g\,m^{-3}$)

c Sternberg's coefficient of attrition (km^{-1})

D hydraulic diameter $= 4\,F/U$ (m)

d grain size (mm, cm)

d_m mean grain size (mm, cm)

d_r effective grain size (mm, cm)

$d*$ sedimentological diameter (mm, cm)

d_F channel frequency (km^{-1})

e_F river development

e_L course development

e_T valley development

F 1. cross-sectional area, flow section (m^2)
      2. force (N)

$\vec{F}$ vector of force (N)

Fe_f solid load (t)

Fr Froude number

Fr_0^1 reduced Froude number in density currents

$Fr*$ Froude number for description of transport of solids ($kg\,s^{-1}$)

F_s effective cross-section for bedload transport (m^2)

G weight (N)

G_f bedload transport ($t\,a^{-1}$)

G_s bedload discharge, dry (t/s)

G_s'' bedload discharge, weighed under water ($t\,s^{-1}$)

G_v bedload attrition (g, kg)

$\vec{G}_f$ vector of gravity (N)

g gravity acceleration

g' reduced gravity acceleration in density currents ($m\,s^{-2}$)

g_s unit bedload discharge after du Boys and others ($t\,ms^{-1}$)

g_s'' unit bedload discharge weighed under water, after Meyer-Peter ($t\,ms^{-1}$)

H height of bedforms (m)

h water depth (m)

h_c critical water depth in density currents (m)

h_{gr} limiting water depth for bedload transport (m)

J energy gradient

J_r friction slope

J_q transverse inclination of water level

J_s bed slope

J_w hydraulic gradient

k equivalent roughness (m)

k_r grain roughness after Strickler ($m^{1/3}\,s^{-1}$)

k_s velocity coefficient after Manning-Strickler ($m^{1/3}\,s^{-1}$)

k_1 attrition coefficient after Stelczer (h^{-1})

k_2 abrasion coefficient after Stelczer ($mm^3\,h^{-1}$)

L length of salt tongue for density currents (m)

l_F river length (m, km)

l_M meander length (m, km)

l_T valley length (m, km)

M meander arc length (m, km)

m mass (g, kg)

n 1. porosity
       2. fragility coefficient after Düll

δ vector of angular velocity (s^{-1})

P bedload particle weight (N)

p hydrostatic pressure (N/m^{-2})

Q discharge ($m^3\,s^{-1}$)

Q_G bank-full discharge ($m^3\,s^{-1}$)

Q_s effective discharge for bedload transport ($m^3\,s^{-1}$)

q discharge per unit width ($m^2\,s^{-1}$)

q_0 discharge per unit width at start of bedload transport ($m^2\,s^{-1}$)

q_T mean transport of solids by bedforms ($m^2\,s^{-1}$)

R hydraulic radius (m)

R_s reduced hydraulic radius after Meyer-Peter and Müller (m)

$\vec{R}$ vector of resultant force (N)

Re Reynolds' number

Re^* Reynolds' number for description of a transportation process of sediments

r radius of curvature

S concentration of suspended load

s height of upstream side of bedforms (m)

t time (s)

U wetted perimeter (m)

u advance velocity of bedform ($m\,s^{-1}$)

u shear stress velocity ($m\,s^{-1}$)

v velocity (m s^{-1})

v_c critical velocity (m s^{-1})

v_m mean velocity (m s^{-1})

v_s flow velocity above bottom (m s^{-1})

v_{rel} relative velocity (m s^{-1})

$\vec{v}_{rcl}$ vector of relative velocity (m s^{-1})

w 1. surge velocity (m s^{-1})

       2. settling velocity of suspended particles (cm s^{-1})

γ specific gravity, in particular of water (N m^{-3})

γ_s specific gravity of solids (N m^{-3})

ε 1. void ratio

       2. relative roughness = k/d

η 1. dynamic viscosity (Ns m^{-2})

       2. relative water depth after Welikanov

Λ length of dunes and antidunes (m)

λ coefficient of resistance, c. of friction

μ coefficient of turbulent friction after Hjulström (cm^2 s^{-1})

ν 1. kinematic viscosity (m^2 s)

       2. gradient reduction coefficient after Putzinger (km^{-1})

ϱ density of water ϱ_w and solids ϱ_s (g cm^{-3})

ϱ' density ratio $(\varrho_s - \varrho_w)/\varrho_0$

τ viscous shear stress (N m^{-2})

τ_0 wall shear stress (N m^{-2})

τ_c critical shear stress (N m^{-2})

τ_h shear stress in density currents (N m^{-2})

τ' apparent shear stress after Prandtl (N m^{-2})

Φ transport intensity after Einstein

φ geographic latitude (^{0})

Ψ intensity of movement after Einstein

ω angular velocity (s^{-1})

1 Introduction

Year by year some $100 \times 10^{12} \, m^3$ of water fall on the earth's surface as precipitation. About two thirds of this is reevaporated, while the remaining about $37 \times 10^{12} \, m^3$ finds its way under the influence of gravity to the sea. When the water does not percolate, the run-off leads to the formation of small rills, while cloud bursts at times cause sheet floods. These are a special type of areal, unconfined run-off. Lower down, the waters feed larger channels which form creeks, streams, and eventually rivers flowing to the oceans, thus maintaining the process of circulation of water. The water retention capacity of natural storage media such as the subsurface and lakes, as well as the vast accumulations of snow and ice allows most rivers − at least in the more humid climatic zones − to maintain their flow even after extended periods of drought. Such perennial streams contrast with those which seasonally or periodically disappear, to resume flow after strong downpours, like the streams of many arid regions.

It is a common characteristic of all types of rivers that they, more than other agencies, mold the land contours of the earth. Wherever a rain drop hits the bare soil, small particles are loosened and made available to be washed away. With the growth of each water course, the transport capacity of the water also increases and with this the extent of river erosion and accumulation. Wherever water flows, there are no stationary stages, be this the process of slow denudation or the short-lived force of a mountain torrent. The differences in elevation on the surface of the earth caused by endogene forces are mainly leveled off by the continuous action of water (except for eolian erosion) until changes in the internal equilibrium or tectonic processes impart a different direction to this process. Thus the land surface has no permanent shape, but is subjected to slow or rapid changes, locally at such a rate that within one generation these changes may be clearly recognized.

It is the purpose of geomorphology to describe and categorize the various configurations of the earth's surface. A special subsector, fluvial or river morphology, deals with the origin and the formational processes of rivers in the widest sense. It is also a descriptive science obtaining its insights mainly from the careful observation of natural phenomena. This discipline also employs the research results of a number of related sciences such as physics and geology, hydraulic engineering and hydrology, climatology and landscape ecology. The many contacts with hydraulic engineering have also resulted in numerous impulses and mutual support. Hydraulic engineering experiments have even been used for solving certain problems of river morphology.

River landscapes appear to the observer as the product of an intricate and unceasing interaction of forces. To reveal their origins it is first necessary to gain an insight into these forces in relation to their natural substrate, i. e., water and rocks. In this context, channel hydraulics are an important aid in attempting a description of the flow processes by physical laws and empirical coefficients. From the wide field of hydraulics only a few points will be dealt with in this book, sufficient to facilitate the understanding of certain kinetic processes and to point the way to further in-depth studies to the planning engineer.

The theory of sediment transport in rivers is closely linked to channel hydraulics. In laborious investigations, this discipline has been able to unravel the laws of movement of solids, bringing many problems of river morphology closer to a solution. This theory has attained importance in modern river management. In view of the vast expanse of related knowledge, it is not possible here to present more than a guideline. Only certain methods of calculation predominantly used in central European practice will be discussed in some detail.

The same constraint also applies to the field of sedimentology, which in recent years has grown by leaps and bounds. Of these results, only the information on grain size distribution and on the composition of the bed load and the suspended load is used to some extent in river morphology. The origin of sediments as a fluvial or limnic product is well known. Likewise, the geoscientist can use structural and textural data of sediments in an actualistic approach in addition to the typical characteristic fossils for differentiating sedimentary rocks of earlier periods into marine and terrestrial, and amongst the latter with some certainty into material of fluvial, limnic, or eolian origin.

As far as the source of the solids is concerned, the hardness of the rock is of prime importance. The hardest and toughest rock will offer the greatest resistance to abrasion and attrition. While accounting at the start of its travel for only a few percent in weight of the bedload, it can eventually become quite significantly enriched, for example at the river mouth. The role of the hardest rock can also be taken over by rocks which at first sight do not qualify as such. Furthermore, in different climatic environments rocks tend to react differently, and consequently the composition of the bedload or the suspended load cannot necessarily be used as an indicator for the geological make-up of its source area.

It is one of the most important objectives of river morphology to describe the geometric parameters of river beds and to interpret their original causes with the aid of basic theoretical knowledge of the kinetics of water flow and of the solids thus transported. As the configurations are extraordinarily multi-faceted and shifting, it is necessary to subdivide them according to their characteristic features. In the individual presentation of the geometric elements of river beds one must not forget that rivers are three-dimensional structures which are subject to a multitude of changes with time.

One of the aims of this book is to stress an all-encompassing view of a river such that the description of the plan view also takes into account the development of the longitudinal and transversal sections. Just as there is a tendency to view the mutual interaction of the living and nonliving constituents of our globe as an "interdependent system", this term also applies to the specific system of a river. A number of empirically derived rules for the calculation of certain factors will complete the presentation of the geometry of channels.

The fact that rivers are not distributed accidentally across the face of the earth, but are rather dependent on climate and rock type, led to early classifications. Attempts were made to characterize and classify the various types. Each type of river mentioned is to some extent morphologically controlled, as its run-off — more or less irregularly distributed over the year — also exerts certain morphological effects. There is thus no "pure" type and usually one morphological pattern dominates over the others, resulting in numerous mixed types.

The history of rivers allows the deciphering of how greatly climates and landscapes have changed over geological periods. These changes are a visible expression of the interaction between uplift, subsidence, and horizontal movements of part of the earth's crust and of the climatic influences. The flowing water can accordingly react either as an erosive or accumulative medium. Here reference can be made to Wundt (1953) "If the river and the surface of the earth here frequently are referred to as if they were acting by themselves ('the river tries to do something, etc.'), this should not be understood literally. There are primary processes (precipitation of rain) and secondary processes (the reaction of the earth's surface to this) which are separated from each other by periods of time. This sequence of events can best be visualized as perseverance, but it has nothing to do with metaphysical forces."

2 Historic Overview

In retrospect, hydrological research and its achievements are usually seen from two different points of view. On the one hand, there is the scientist who emphasizes geophysical, geological, and geographic data and interpretations. On the other hand, there is the engineer who recognizes in the water supply and irrigation systems of ancient peoples a great wealth of experience in the management of water resources. Additionally, he will turn his attention to the measuring techniques which, at first developing rather slowly, have been improving rapidly from the start of modern times.

The hunters and gatherers invariably located their dwellings close to waterholes or other open sheets of water, also because the better hunting grounds were to be found there. With the dawning of the early high cultures, i.e., when larger population groups decided to settle and cease migrating, the observation of waters and of water-related phenomena began. As Kreps (1967) commented: "the observation of hydrological elements increased in importance and led to the age-old recognition that the amount of natural precipitation, the water level of a river, and the irrigation of large areas fed by this river determined the success or failure of most harvests. The impressive hydraulic engineering achievements of prehistory and of the early classical periods in the Middle East and Mesopotamia, Egypt and Sri Lanka, India and ancient China, Java and the Inca states of South America would not have been possible without some basic hydrological observations. We know now that ...Sumerian science was already in a position to calculate the date of the start of the rainy season and the approach of the melt water in the river. The power of the Mesopotamian rulers was based, as it were, on their effective control of the floods of water brought down by the rivers. The regulation and the distribution of the water proved the basis for the whole state. This would have been unthinkable without hydrological observations and evaluations...".

Written records are sparse; only in Egypt are there some clues giving information on the varying water level of the Nile. The Palermo stone tablet (2800 B.C.), which reports the water levels of the Nile over a period of 52 years, as quoted by Kresser (1967), represents the first known hydrological publication. However, there are virtually no commentaries or interpretations, although the livelihood of the people was governed by the regular seasonal changes of the river level and the river was everybody's concern. More extensive written records are available only from Greek times onwards, from philosophers as well as from geographers. The synopsis, the integral view of the relevant phenomena, was more in tune with Greek speculative thought than with the analytical thinking of our times. In earlier times all phenomena perceived were shrouded in myths. In their attempt to elucidate the causal connections, the Greek philosophers did not move as far from the mythical foundations as the unprejudiced methodology of any present-day science would demand, inspired by Galilei's motto: "measure what can be measured, make measurable what cannot be measured." For the ancients our analytical way of thought was less important than harmonizing the obvious multitude of natural phenomena into one grand concept.

The concept of water circulation, however, was by no means alien to the Greeks, and as also in later periods, they referred to the analogy with the human blood circulation. However, the complicated run-off conditions in the karst-affected Mediterranean regions did not favor an easy assessment of the water supply situation, but proved rather to be an impediment for an early clarification of the respective relationships.

For the same reason, the preparation of detailed maps was not considered important, although there were a number of early attempts at a small-scale representation of the world as known at the time. "The fact that, despite some promising starts, the development of the study (of the natural phenomena) was stifled for more than a millennium has some obvious roots in the development of the classical period itself, and not in the decline of science during the Middle Ages. Although there was no lack of methods, which frequently were comparative and causal at the same time as well as being based on observations, there was a notable lack of any attempt to unveil geological structures such as the concept of development. The geographical experiences of Hekataios, Herodotos, Phytheas, Polybios, and Strabo must not be underrated. But there was no map available, . . . there was not even the possibility to represent the landscape in a corresponding way" (Maull 1958).

The totality of scientific and cosmographic knowledge was compiled by Aristoteles (mainly for meteorology), Strabo (in his *Geography*), Pliny the Elder (*Historia Naturalis*), and Seneca (*Naturales Questiones*). Of these, Seneca can be considered as the strictest systematician.

On the whole there was preserved a multitude of sometimes astonishingly exact observations of processes and phenomena which affected the human condition or which simply appeared memorable. In addition to volcanic eruptions and earthquakes, the activity of rivers and the changes of the coast lines were recorded as the most obvious phenomena. Far-reaching changes of the coast line were caused by rivers which gradually led to the silting-up of whole bays and to the decline of famous trading ports such as Miletos, Ephesos, and Spina (cf. Chap. 7).

According to the old philosophers, the circulation of water took place through subaerial and subterraneously linked river systems. What we now call the atmospheric circulation of water was not yet developed as a mental concept, despite some indications to this effect. Plato was of the opinion that in addition to liquid fire and flows of mud and air in the earth's interior, water also kept seeping or pouring underground, thus restlessly migrating around. The earth's interior was considered to be full of caves, pipes, channels, and narrow veins, and so the water was thought to penetrate along these to the center of the earth, but not beyond, and consequently had to undergo complicated movements. Being forced back partly to the surface, it fed rivers and lakes in the form of springs (*Phaidon*).

Aristotle is considered as the founder of the condensation theory, which held its ground for a considerable time and was replaced only rather late in history: rain water seeping down gathers in cavities (just as did air, which actually was visualized as finely disseminated water), and from there reappears in wells and springs. The widely developed karst hydrography led to a variety of erroneous interpretations of the permeability of rocks in general. Several rivers were thought to reappear at different localities. Pausanias reports on the river Maiandros; Nile and Niger were thought to be somehow connected, and the Nile supposedly passed through long stretches of ground subterraneously, only resurfacing in Egypt (Pliny). A first criticism of these concepts was voiced by Strabo and Seneca.

The circulating Oceanos was not visualized as the world-encircling sea but rather as a world-encircling river outlining the circumference of the earth's disc. Our term "ocean" thus actually was derived from flowing waters.

As the bulk of the observations was derived from the morphologically highly diverse Mediterranean region, it is not surprising that most speculations about the origin and course of the large rivers in the north and northeast of Europe were essentially wrong. The Danube was known, as well as the lower reaches of the Dnjestr, Dnjepr, Don, and possibly also of the Volga, but it was assumed that they originated in high mountain areas such as one knew from the rivers back home (cf. for example Aristotle) or flowed from lakes of various sizes (Herodotos). The origin of large river systems in plains and tablelands was not yet an easily visualized concept.

In the presentation of the earth as prepared by Erathostenes, the Riphaean Mountains are depicted, a mountain range extending from the Oceanos in the west to the legendary country of the Scythians in the east. This was considered as the great watershed which eventually was sought more in the interior of eastern Europe, as during Roman times the central European geography became better known. In this concept there was undoubtedly an intermingling of the reports on the location and extent of the Pyrenees, Alps, Carpathians, and the Ural, which only gradually became unravelled. The existence of a "mountain divide" somewhere in the expanses of the forests and steppes of eastern Europe was eventually abandoned during the 16th century when the north-south orientation of the Ural became known. The term "montes riphaei" given to the legendary mountain range was then transferred to the Ural. As "Riphaean" the term survives in modern geohistory, where in Russian usage it designates the youngest part of the Precambrian.

With the fall of the West Roman Empire, the work of the Greek philosophers and cosmographers fell into oblivion in the west. They achieved acceptance again only during the Middle Ages via the Islamic scientists of Spain, and were then incorporated in scholastic philosophy. Concurrently with the acceptance of Greek thought, a process of secularization was started which has continued, almost without interruption, into our present times. The teachings of these authorities were not merely taken over, but they were used as a foundation for the independence of scientific research, then evolving. Experimentation became accepted from the 13th century as a second path to knowledge, in addition to pure argumentation. From the peak of the Renaissance onward, the ground had been prepared for a physicomathematical approach to natural phenomena. However, the old deeply rooted convictions were able to survive for quite some time.

During the Renaissance one individual, Leonardo da Vinci, proved so far ahead of his times, that his views and works were for a long time disregarded, and his ideas were even replaced again by false theories. According to Reindl (1939), the results of his hydraulic and geological-geographical observations represent the highlights of his scientific research. His sketches of the movement of water are numerous, and he made remarkably accurate drawings of eddy formation and of the flow around and across obstacles. These sketches and drawings, especially those of the deposits of gravel in bends or braided channels of rivers confirm his exceptional clarity of observation and interpretation.

As a side effect of the continuing parallel development of speculative thought in addition to precise observations of nature and theoretical interpretations of these phenomena, many scientists attempted lucid and bold explanations. Some of these now appear rather exotic to us, but with the dawning of our modern era, cosmographic systems of a type unknown during the Middle Ages were proposed. One of the most wide-ranging of these systems was developed by B. Varen in his *Geographia generalis*, Amsterdam (1650), in which he restricted himself mainly to facts, avoiding all the bolder interpretations. In effect it was Varen who introduced a true geomorphology and hydrology (Maull 1958).

In its time the *Mundus subterraneus...*, *Amsterdam 1665* of A. Kircher was also famous (Scheurmann 1977). Like Varen, Kircher is rooted in the classical period, especially with regard to the description of cavities in the earth's interior (Hydrophylacia). On the other hand, in certain fields such as the observation of the forces of erosion, he went beyond Varen. Kircher was the first to differentiate between regions of erosion and deposition in stream beds.

In a multitude of other treatises, time and again references to the action of rivers can be found. Special emphasis was placed on the phenomena of meanders which already occupied Leonardo and Galilei, and which were treated later by Kant in his physical geography. With the advance of geological science during the 18th century, morphological investigations were frequently carried out, albeit with varying success. As these researches were essentially morphographical, i.e., descriptive and measuring, until far into the 19th century, geomorphology, and with it river morphology, in so far as it was executed geographically, remained rooted in the subject of geography. Only with the increase of morphogenetic methods, especially in connection with climatic geomorphology, did geological aspects find more recognition in geomorphology and vice versa.

With the start of the first major river regulations in the 19th century, the engineers in charge found themselves faced with morphological problems of an extent hitherto unknown. As one example for numerous others, the course correction of the Rhine below Basel by Tulla can be mentioned. With the attempt to master the hydraulic and sedimentation problems, river morphology found a new place among the engineering sciences. Depending on the point of view, it has since become customary to consider river morphology mainly under morphogenetic or hydraulic engineering aspects.

3 Hydraulics of Stream Flow

3.1 Basic Laws of Channel Hydraulics

3.1.1 Steady Uniform Flow

The calculation of flow velocity is one of the most important objectives of channel hydraulics. The common methods are based on the assumption of a uniform velocity developed across the complete channel cross-section. From the equilibrium conditions of Fig. 3.1 at a hydraulic diameter of $D = 4R = 4F/U$, the wall shear stress or tractive force will be

$$\tau_0 = \tfrac{1}{4}\varrho_w g JD \ . \tag{1}$$

The velocity of the shear stress is found from

$$u_* = \sqrt{\tau_0/\varrho} = \tfrac{1}{2}\sqrt{g JD} = v\sqrt{\lambda/8} \ . \tag{2}$$

The general law of flow in channels is

$$v = \frac{1}{\sqrt{\lambda}}\sqrt{2g JD} \ . \tag{3}$$

As in rivers, flow is almost exclusively of the turbulent type, the resistance coefficient λ depending not only on the Reynolds number $Re = vD/v$ but also on the relative roughness coefficient $= k/D$ of the channel wall. For calculations in the so-called transition zone between "hydraulically smooth" ($\varepsilon = 0$) and "completely rough" ($Re \rightarrow \infty$) the transition law of Colebrook-White applies (Press-Schröder 1966):

$$\frac{1}{\sqrt{\lambda}} = -2\log\left[\frac{2.51}{Re\sqrt{\lambda}} + \frac{k/D}{3.71}\right] \ . \tag{4}$$

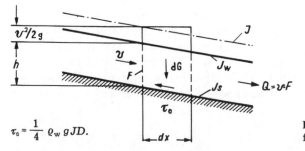

$$\tau_0 = \frac{1}{4}\varrho_w g JD.$$

Fig. 3.1. Slope of energy line at uniform steady flow

The application of this equation without the use of nomograms is rather cumbersome, as λ cannot be closely calculated. The approach of Martinec (1967) is more suitable for the practical use of the mean flow velocity:

$$v_m = \left(17.7 \, \log\frac{R}{d_{50}} + 13.6 \right) \sqrt{RJ} \; . \tag{5}$$

It represents the result of numerous measurements and is based on the theory of a logarithmic distribution of velocities (Sect. 3.1.2). For rivers with highly variable cross-sections a correction is required, as the change between large and small cross-sections continuously leads to local accelerations and retardations which result in losses of energy. In other words: the larger the ratio F_{max}/F_{min} becomes, the larger the flow resistance will be. The determined grain size, d_{50}, in this case is replaced by the fictitious grain size, d_n, which incorporates the influences of bed roughness and the various irregularities of the river channel. Based on experiments there is

$$d_n = d_{50} + \Delta d \; , \tag{6}$$

$$\Delta d = 0.263 \frac{F_{max}}{F_{min}} - 0.322 \; . \tag{7}$$

Irregular shapes of the bed must be assumed when F_{max}/F_{min} is above 1.2. Above a ratio of 2.1 the method can no longer be applied, as with still higher variability of the cross-section the presence of nonuniform flow conditions can no longer be neglected.

When determining F_{max}/F_{min} a suitable length of the river course and proper distances between the cross-sections measured have to be chosen. Experience has shown that the length should be equivalent to 20–40 times the width of the river, and that the interval between sections should be such that there are at least 40 of them over the full length investigated.

In addition to the method described above, for quite some time empirical flow formulas have been in use. Amongst the so-called potential equations of the type

$$v = \text{number} \times R^\alpha J^\beta$$

in practical work mainly the Manning-Strickler law with $\alpha = 2/3$ and $\beta = 1/2$ is applied. It thus reads

$$v = k_s R^{2/3} J^{1/2} \; . \tag{8}$$

As shown by detailed investigations, the exponents α and β strictly speaking are not constants, but depend on the properties of the bed (Bogárdi 1974). It has become customary, however, to use them as constants. At grain diameters $d_m = 3$ mm and 10 mm at the bottom, the experimentally derived exponents are in any case in good agreement with the Strickler numbers.

The correct selection of the velocity coefficient k_s always presents some difficulties as it is influenced by the roughness of the channel wall, and additionally by all peculiarities of the river bed, such as irregular cross-sections, losses in bends, and the influence of the bedload transport. For the true grain roughness

Table 1. Approximation of k_s for various river bottoms.
(Meyer-Peter and Lichtenhahn 1963)

Properties of river bed	k_s
Soil profile, regular, no bed load	40
Coarse gravel material, regular	37
Rivers with low bed load, coarse gravel	33 – 35
Irregular river beds with gravel	30
Mountain rivers with coarse gravel, bedload stationary	25 – 28
As above, bedload in motion	19 – 22

on a plane bottom with rigid walls at fully developed turbulence Strickler (1923) has developed from experiments the formula

$$k_r = \frac{26}{d_r^{1/6}} \ . \tag{9}$$

The critical grain diameter d_r roughly corresponds to the d_{90} of the cover layer of the bottom. For calculating the velocity coefficient k_s which results from the shape roughness, Eq. (9) has to be modified to

$$k_s = \frac{K}{d_r^{1/6}} \ . \tag{10}$$

Due to the always present irregularities of natural rivers the value for K is always below 26. Strickler himself gives an average of $K = 21.1$, whereas according to the evaluation of a larger number of results by Garbrecht (1961), it ranges between 16 and 26. According to Müller (1960), the ratio k_s/k_r decreases from 1.0 for a plane bottom to 0.5 for high ripples. For the calculation of the constant K the experimental results of Meyer-Peter and Lichtenhahn (1963) can be applied. Figures for an approximation of k_s are presented in Table 1.

From recent literature it is known that Eqs. (59) and (60) result in excessive values for the flow velocity in bedload-bearing rivers, especially for a large relative roughness. From experiments Jäggi (1984) developed an improved numerical equation:

$$v_m = 2.5 \, u^* [1 - \exp(-0.02 Z / J^{0.5})]^{0.5} \ln (6.1 Z) \ , \tag{11}$$

with the relative roughness $Z = h / d_{90}$. From Eq. (11) the Strickler coefficient K of Eq. (10) can be recalculated. For $Z > 10$ values of $20 < K > 22$ result, whereas K can drop significantly for $Z < 10$. Under these conditions Strickler's Eq. (8) should not be applied.

3.1.2 Distribution of Velocities

When calculating the mean flow velocity, no statements are made with regard to the distribution of the velocity between the water surface and the river bed. In earlier times this problem was approached through empirical considerations. Rinsum (1950), for example, used the quadrant of an ellipsoid to represent the vertical

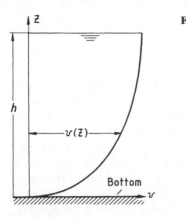

Fig. 3.2. Vertical distribution of velocities

velocity distribution. Modern authors base their works more on Prandtl's boundary layer theory. According to this, even under highly turbulent conditions, there is a thin layer directly along the walls within which laminar flow according to the Hagen-Poiseuille law is developed. Using the auxiliary term "mixed path" and concepts of the exchange of impulses, Prandtl arrived at a logarithmic law of velocity distribution. According to Fig. 3.2, in the so-called hydraulically rough region there is

$$\frac{v}{u_*} = 8.48 + 5.75 \log \frac{z}{k} \, . \tag{12}$$

Unbehauen (1970) has shown that the logarithmic law of distribution is universally applicable. According to his investigations, the wall roughness K of the river bed can be approximated by the mean grain size d_m of the cover layer. A number of correction factors are used for the calculation of the actual effective wall roughness.

The velocity is distributed irregularly not only in the vertical but also in the horizontal plane. In straight and uniform sections of the river course the velocity v_R closer to the sides is somewhat lower than v_A along the river axis.

At right angles to the flow there is thus a drop of pressure

$$\frac{\Delta p}{\varrho} = \frac{1}{2}(v_A^2 - v_R^2) \, , \tag{13}$$

resulting in a slight concave upward curvature of the water surface. This leads to a secondary current directed inward which transports floating material from the sides into the central axis of the river.

A similar, but much more pronounced phenomenon takes place during rapid surges of wide rivers in flood. It is well established that the surge travels downstream with an accompanying continuous change of the water surface at the velocity

$$\omega = \frac{\delta^2 Q}{\delta t^2} : \frac{\delta^2 F}{\delta t^2} \, , \tag{13a}$$

or approximately

$$\omega = \frac{3}{2}v \ . \tag{13b}$$

It follows from Eq. (13b), according to Forchheimer (1930), that the surge will attempt to advance more rapidly where the flow velocity is high, i.e., it will move ahead along the line of highest flow velocity. During the rise of the flood this results in a convex curvature of the water level which at times can be rather impressive. On the Po River at Causale Monferrato in 1857 a rise of 1.5 m was observed over a width of 300 m. On the Loire in 1866 at 165 m width there was even a rise of 2.4 m. The opposite effect can be observed after the passage of the surge wave, albeit in a much weaker form due to the slower fall of the flood. According to Samojlov (1956), the vertical rise (or fall) of the water level is a function of the transverse gradient of the bed from the sides to the median axis of the river and of the velocity of the rise or fall of the water. The water level will bulge stronger at right angles to the flow direction, the shallower the gradient of the bed is in the same direction.

3.1.3 Flow in Meandering Channels

Natural rivers are usually straight only over short distances. Their plan view is generally made up of a sequence of bends. The current is then subject to flow conditions other than in straight sections. Over regular bed forms and under depths which are great when compared to the width, the current in bends will accelerate on the inside bank and slow down on the outside bank (Fig. 3.3).

According to the so-called law of angular momentum, which is strictly applicable for potential currents only, the velocity in circular bends is hyperbolically distributed across the width according to $v = C/r$. On the outside bank the water level will rise, whereas it will fall on the inside bank.

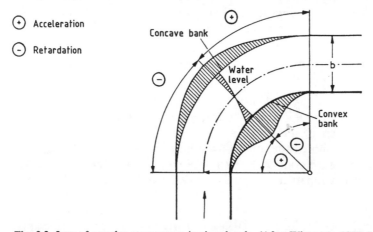

Fig. 3.3. Law of angular momentum in river bends. (After Wittmann 1955a)

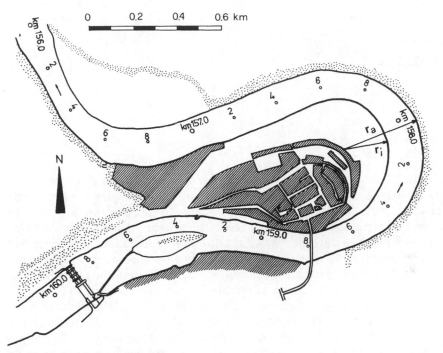

Fig. 3.4. Inn meander at Wasserburg

When the depth-width ratio, as in the majority of cases, is rather low, the influence of friction calls for a different approach. In river bends with the radius r the centrifugal force causes a transverse slope of the water level of

$$J_q = \frac{1}{grh} \int_{z=o}^{z=h} v^2 dz \; , \tag{14}$$

or approximately

$$J_q = \frac{v_m^2}{gr} \; . \tag{15}$$

For rectangular cross-sections the differences in water level between inside and outside bank can be approximated with the aid of the formula

$$\Delta h = \frac{v^2}{g} \ln \frac{r_a}{r_i} \; , \tag{16}$$

where r_a and r_i are the corresponding curvature radii.

To illustrate this with an example, the Inn flood of August 7th, 1985 in the Wasserburg meander may be considered (Fig. 3.4). During the passage of the wave crest, at the center of the bend the difference in water level between the inside and the outside bank amounted to $h = 0.55$ m. Registering a mean flow velocity of $v = 3.0$ m s^{-1}, with $r_a = 350$ m and $r_i = 200$ m, Eq. (16) results in:

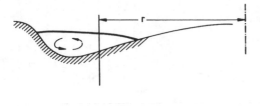

Fig. 3.5. Twisted flow in river bends

$$\Delta h = \frac{3^2}{g} \ln \frac{350}{200} = 0.51 \text{ m} \ .$$

This result differs only slightly from the observed difference in elevation.

On the outside bank of the curve there is thus a greater pressure head than on the inside bank. However, as shown in Section 3.1.2 above, the flow velocity in the vertical section according to the law of logarithmic distribution is highest near the water surface and lowest near the bottom. As a result the higher-up water particles are exposed to higher centrifugal forces than the deeper ones. Consequently, the transverse gradient, which can be calculated from the integral of the velocity as a function of the distance from the bottom (z) according to Eq. (14a), can no longer keep the individual water elements in equilibrium. In order to compensate this pressure differential, the flow close to the surface at the outside bank goes down and displaces the water present there to the inside bank. In connection with the downstream flow a helically twisted flow results, which tends to erode the outside bank and will lead to deposition on the opposing bank (Fig. 3.5). In general, however, the twisted flow does not form a completely closed roll, but will be restricted to the zones close to the side, allowing only incomplete mixing of the waters in the bends. Under certain conditions even two rolls with counter-rotation can develop in wide lowland rivers.

The kinetic energy of the helical current roll according to Wittmann (1955) amounts to about $2-3\%$ of the total flow energy. Its proportion is reduced as the ratio r/b grows. A decrease of the ratio r/b also results in a decrease of the kinetic energy. On the other hand, it is increased as the angle of curvature grows.

Energy losses in sinuous river stretches can be calculated according to Müller (1943) or Garbrecht (1961). It is, however, more convenient to use empirical approximations. Wittmann recommends either to use Strickler's k_S value or to reduce the calculatory slope J against that of straight sections of equal length. At an r/b of about 5 and $\alpha = 60-135°$ the value for k_s can be reduced by about 20%, whereas J will be lower by $5-10\%$ than for the straight section.

3.2 Influence of the Earth's Rotation

3.2.1 Baer's Law

The older geographic literature pays much attention to the question of the influence of the earth's rotation on the movements on its surface. Up to the

mid-19th century the standpoints of Hadley (mentioned in Henkel 1922) prevailed, who ascribed the direction of the trade winds to the rotation of the earth, using the difference in velocity at different latitudes as the explanation. It was then
gradually recognized that every movement of a body on the earth is deflected, to
the right on the northern and to the left on the southern hemisphere. However,
it was assumed that the theoretically postulated deflection was too weak to exert
any notable influence on flowing water.

In 1860 the Baltic scientist K.E. von Baer published a report in the Annals
of the St. Petersburg Academy of Sciences under the title: *On a general law of
formation of river beds*. He attempted to prove that a fact that had been known
for a long time, but had been rather enigmatic, i.e., the steep nature of the right
and the flat shape of the left bank of most major rivers, could be explained
through the rotation of the earth. This long and hotly debated theory has since
become known as Baer's law. The introductory statements of the author himself
are: "The flowing water, coming from the equator to the poles, will have a rotational velocity greater than that of the higher latitudes and thus will press against
the eastern bank, as the rotation is directed east and with it also this little surplus
brought along by the water flowing from lower to higher latitudes. On the other
hand, water flowing from more or less polar regions to the equator will arrive
there with lower rotational velocities and thus press against the western banks. In
the northern hemisphere for rivers flowing north, the east bank is the right one,
whereas for rivers flowing south, the west bank is the right one."

As proof, the author described his observation of a number of rivers and
especially of the Volga, the right bank of which in Russian is called the "hill
bank" and the left bank the "meadow bank". The high hill bank is the preferred
area for settlements, whereas in the swampy flats on the left bank only few
villages are found. The hill bank is subjected to continuous erosion which result
in the collapse of enormous masses of rock from time to time. Olearius described
in his *Persian Travel Notes* (1666), that this process led to the burial of an entire
ship and its crew which had been lying in the shelter of the bank above Astrachan.

Baer collected reports not only on the Russian rivers but also on the other
rivers of the earth such as the Rhine, Nile, Mississippi, la Plata, etc. With few exceptions, for example the left-hand turn of the Euphrates near Aleppo, he found
support for his law of dextral deflection. He stated that this law would become
more pronounced with a more north-south orientated course of the river. To support his theory he suggested evaluating statistically whether derailments of trains
traveling exactly north or south were more frequent to the right or left. He eventually tried to develop his law purely from physical factors. He was only partly
successful – being himself a biologist – and he had to accept that the forces of
deflection are very small indeed. It was held against him that the real reasons for
the undeniable asymmetry of river beds should be sought rather in the prevailing
wind directions and the condition of the bedrock. Nevertheless, the author
adhered to his convictions with his characteristic intuition.

In the following decades Baer's law was repeatedly attacked. The protagonist
of this opposition was Zöppritz (in Henkel 1922 and Wegener 1925), who criticized Baer severely by stating that in a straight river course the deflecting force
of the earth's rotation would lead only to a gradient of the water level at right

angles to the direction of flow. The difference in height between the right and left water levels would be so small that it could not exert any notable influence. He added: "For the situation on the Siberian rivers which Baer without doubt described correctly, different explanations have to be found. They must be sought with certainty in the westerly winds prevailing there throughout the year".

Fabre (1903) devoted a detailed discussion to the reasons for or against the validity of Baer's law. Based on a 1789 article of the French hydrographer Lamblardie, the *Morphology of the Earth's Surface* of Penck (1894), and his own observation, especially in the Gascogne area of France, he concluded that Baer's proposals on the influence of the terrestrial rotation on the development of the cross-section of rivers could not be upheld. For an explanation of this situation, meteorological and geological factors should be preferred. Fabre's ideas culminated in the statement that the asymmetry of valleys and the deflection of the thalweg are caused by geological and geographic factors, mainly torrent erosion and wind deflation.

During the 1920's, the discussion on Baer's law was revived. Reference must be made here primarily to Henkel (1922, 1928), who submitted cogent arguments for its applicability. Wegener (1925) also arrived at a positive verdict, attacking Zöppritz for insufficient consideration of the dynamics of transport of solids in rivers. Eventually, Exner (1927) found some basic proof for the applicability of Baer's law when allowing water in a test model to flow over a sandy surface on a rotating disc. As was to be expected, the right bank of the channels showed bulges when the disc rotated counterclockwise. Exner mentioned that these tests were also carried out specifically to disperse the doubts on the validity of Baer's law voiced previously by Schmidt (1924).

What is the position of modern hydrology with respect to Baer's law? This can only be answered if the whole interaction of forces in a river is taken into account according to mechanical principles. This will be done in the next two chapters, which are largely based on a paper by Dantscher (1942).

3.2.2 Coriolis Force

According to the principles of relative movement in a rotating system a moving mass is subjected to the so-called Coriolis force which is expressed as twice the vector product of the angular velocity and of the relative velocity of the center of this mass, i.e.,

$$\vec{a}_c = 2(\vec{o} \times \vec{v}_{rel}) \ . \tag{17}$$

It amounts to

$$a_c = 2\omega v_{rel} \sin\varphi \ , \tag{18}$$

if φ is the angle between the two vectors. In our case the rotating system is the earth itself, the water of a river being the body in relative movement to it.

When placing a tangential plane E at point A of the earth's surface (Fig. 3.6), it can be stated with sufficient accuracy that the velocity vector of a river at A will always fall into this plane, notwithstanding in which direction it is flowing.

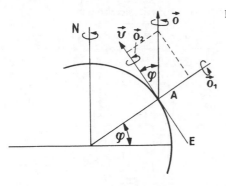

Fig. 3.6. Velocity vectors in A

The vector of the angular velocity can be split up into the vector $\vec{o}_1$ at right angles to the tangential plane and a vector $\vec{o}_2$ in the plane itself. The latter does not have to be considered any further. As Fig. 3.6 shows, there is

$$\vec{o}_1 = \vec{o}_2 \sin\varphi ,$$

i.e., the earth rotates at the angular velocity o_1 around a vertical axis in A. As $\vec{o}_1$ and $\vec{v}_{rel}$ are vertical to each other irrespective of the direction of $\vec{v}_{rel}$, the Coriolis force can be calculated according to Eq. (18) for all cases, whether the river flows north, east, south, or west.

For an astronomic day the angular velocity amounts to

$$\omega = \frac{2\pi}{86164} = 0.7292 \times 10^{-4} [\text{s}^{-1}] ,$$

a very small figure indeed. Combining it with a mean flow velocity of 1 m s^{-1}, different Coriolis forces are obtained (Table 2).

However, care has to be taken not to underestimate very small but continuously acting forces. This becomes evident immediately when, according to Henkel (1922, 1928), the Coriolis force is compared to the downstream component of the gravity acceleration according to

$$\text{tg}\,\delta = \frac{2\omega v \sin\varphi}{g \sin\alpha} . \tag{19}$$

Always assuming 1 m s^{-1} flow velocity, this will result in the values shown in Table 3.

Table 2. Coriolis force at different geographic latitudes

Latitude (deg)	Region	$\sin\varphi$	$a_c \times 10^4$
0	Sources of the Nile	–	–
30	Apex of Nile delta	0.500	0.73
45	Po plains	0.707	1.03
60	Siberian rivers	0.866	1.26

Table 3. Coriolis force compared to downstream component of gravity acceleration

Latitude (deg)	River	$\sin \alpha \times 10^3$	$\sin \varphi$	$\operatorname{tg} \delta$
45	Po at apex of delta	0.07	0.707	0.150
48	Isar below Munich	2.0	0.743	0.005
49	Danube below Regensburg	0.43	0.753	0.026
53	Elbe below entry of Havel	0.13	0.800	0.091
53	Volga at Kujbychev	0.03	0.800	0.396

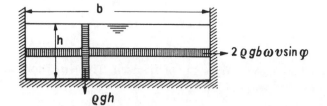

Fig. 3.7. Bottom pressure and Coriolis force in a wide rectangular channel. (After Dantscher 1942)

As Table 3 shows, the values for $\operatorname{tg}\delta$ are highly variable. This fact led Henkel to comment: "The effect of Baer's law is very important for rivers of rather small gradient. At intermediate gradients it becomes indistinct due to local influences and it disappears in mountain streams. The boundaries have to be defined by observation."

Dantscher supplemented these ideas by the following thought: "The unit area of the bottom of a rectangular channel according to Fig. (3.7 in this book) is subjected to the water pressure $\varrho g h$, whereas the right-hand wall in addition to the hydrostatic pressure also suffers the effects of the Coriolis force".

The latter increases with the width of the river. On the other hand, in rivers that have cut their beds into their own alluvial sediments, there is some connection between width, depth, and gradient. For width and depth Siedek (according to Dantscher) has found empirically:

$$h = \sqrt{0.0175\,b - 0.0125} \ . \tag{20}$$

Accordingly a river of 1000 m width will be 4.2 m deep. Assuming a flow velocity at the passage of a flood peak of 3 m s^{-1} and a latitude of 45°, the ratio between bottom pressure and the wall pressure resulting from the Coriolis force will be

$$\frac{gh}{2b\omega v\sin\varphi} = \frac{9.81 \times 4.2 \times 104}{2 \times 1000 \times 0.7292 \times 3 \times 0.707} = 123.4 \ .$$

A stronger erosion of the right-hand bank thus must be considered at least as a possibility.

According to Dantscher it is dubious to postulate a functional relation between the degree of bank erosion and the size of the wall pressure, following, for example, the concept of shear stress and bedload abrasion. However, he refers to a note by Penck saying that in the Rhine between Strassburg and Maxau, as well as in the regulated Danube near Vienna, the pools on the right-hand banks are usually deeper than those on the left-hand bank. In river furcations in a delta the right arm should carry more water than the left arm, unless counteracted by other factors. This is the case for example for the Nogat branching off from the Vistula (Dantscher, Samjolov).

3.2.3 Centrifugal Force

In addition to the Coriolis force there are two other forces resulting from terrestrial rotation. The first one will be referred to only briefly.

According to Fig. 3.6 the component of the angular velocity falling into the tangential plane E is

$$\vec{o}_2 = \vec{o}\cos\varphi \ .$$

During the rotation of the earth the tangential plane rotates around $\vec{o}_2$. It sinks in the east and rises in the west. A body moving east on a certain latitude then appears to lose weight whereas a body moving west appears to gain weight. For a river this would imply that when flowing east it will form a smaller gradient than in the opposite direction. The effective forces, however, are so exceedingly small that their morphological influence can be neglected.

The second or the centrifugal force is much better known and more important.

The centrifugal force (Fig. 3.8) effective at A is

$$a_z = r\omega^2\cos\varphi \ . \tag{21a}$$

It can be split up into the components

$$a_{z1} = r\omega^2\cos^2\varphi \tag{21b}$$

and

$$a_{z2} = r\omega^2\cos\varphi\sin\varphi \ . \tag{21c}$$

Component a_{z1} is opposed to the acceleration by the mass of the earth and reduces the latter most at the equator and least at the poles. Component a_{z2} attempts to pull every mass to the equator. It causes the flattening of the geoid and attains its maximum at $\varphi = 45°$. As the earth's shape deviates from a true sphere, the force resulting from g and a_z according to Fig. 3.9 will not pass exactly through the center of the earth. The normal plane in A will be at an angle to the tangential plane.

The level of a stationary body of water will follow the shape of the geoid. In a north-south-flowing river the fact that the plane at right angles to the resultant force is less inclined that the horizontal plane will result in a southerly acceleration. According to Eq. (21c), it will amount for example to about 0.017 m s^{-2} at 45° latitude, i.e., about 1.73% of the gravity force. Its effect is an increase of the gradient resulting in an increase of velocity. In a south-north course the acceleration tends to decrease the gradient. There are, however, no substantiated observations on the bed-forming effects of the centrifugal force. Dantscher's assumption

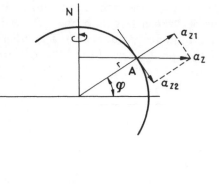

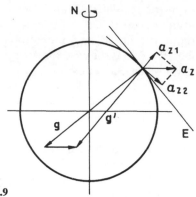

Fig. 3.8   **Fig. 3.9**

Fig. 3.8. Centrifugal force in A

Fig. 3.9. Force acting on geoid. (After Dantscher 1942)

that the strong gradient of some south-flowing Alpine rivers such as the Ticino and the Rhône could be ascribed to this cannot be supported. The reason here can be found rather in the tectonic structure of the Alps and in thrust movements directed mainly northwest to north. They are especially strong over the former area and according to modern geophysical investigations are connected with the pronounced subsidence of the Po plains. Furthermore, the Ticino and the Rhône cannot be compared with each other.

3.2.4   Influence on Tidal Rivers

Tidal rivers differ from nontidal rivers especially in the regular reversal of the flow direction under the periodicity of the tides up to the boundary of the tidal flow. Each bend will then at times describe a right-hand curve and at other times a left-hand curve. Under the influence of the Coriolis force the main current will thus alternate between the two banks. In the middle of the river lower velocities will prevail, favoring the formation of long sand banks when the bed is of sufficient width.

According to Hensen (1939) the direction of the river flow can be presented in so-called "Ersatz" or "equivalent" curves. In calculation the radius of such a curve is obtained from the condition that the centrifugal force effective in it must be equal to the sum of the dextral deflection and the centrifugal force in the actually developed curve. The equivalent curve is visualized as a combination of the effects of the dextral deflection and the natural configuration of the river bed.

In the formation of a delta mouth the Coriolis force, according to Samjolov, is of great importance. The dextral deflection caused by it results in helically twisted water movement transporting the bedload to the center of the river. Directly where the river enters the sea, the bottom will be scoured in a wide depression along the river axis. Behind this a central shoal is formed which necessarily splits the river from a certain stage of development onward into a right and a left arm. The general development of a delta will be discussed in more detail in Chapter 6.3.2.5.

3.3 Density Currents

At the entry of rivers into lakes or the sea the phenomenon of so-called density currents can be observed. It results when waters of differing densities come into contact with each other. Such differences in density can be caused by different temperatures, by the suspension load of the river, or the salt content of the sea into which the river flows. We discuss here the frequently observed case of a planar density current entering a basin, overlaid by a stationary mass of water (Fig. 3.10).

According to Press and Schröder (1966), the density current will remain intact under uniform flow conditions at a thickness h, as long as the flow is laminar. On change-over to turbulent flow conditions, the density flow and the overlying water layer will start to mix through mass transfer, leading to a loss of uniformity of the inflowing water body. From equilibrium considerations which are beyond the scope of this book, the shear stress along the bottom will be

$$\tau_0 = \frac{1}{4} \varrho g \left(1 - \frac{\varrho_w}{\varrho}\right) JD \ . \tag{22}$$

Save for the values in brackets this equation is identical to Eq. (1). In analogy to Eq. (3), the general law of flow is

$$v = \frac{1}{\sqrt{\lambda}} \sqrt{2g'DJ} \ , \tag{23}$$

where $g' = (1 - \varrho_w/\varrho)g$ is the reduced gravity force. With some deviation from the general theory of laminar channel flow the coefficient of resistance $\lambda = 86/Re$, the Reynolds number being $Re = vD/\nu$.

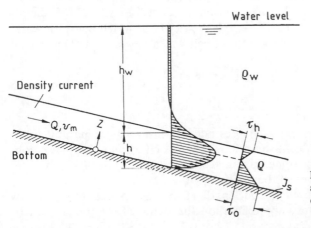

Fig. 3.10. Distribution and boundary shear stress in uniform density current. (After Press and Schröder 1966)

The mean velocity of the density current can be found with the aid of the Napier-Stokes equation of movement. With the experimentally found ratio of the wall shear stress $\tau_h/\tau_0 = 0.64$ for smaller density differences the mean flow velocity of the planar density current will be:

$$v = 0.138 \frac{g' J h^2}{v} = 0.138 \frac{\varrho - \varrho_w}{\eta} g J h^2 \ . \tag{24}$$

The hydraulic diameter in this case is $D = 2.44 \, h$.

The formation of a density current on entry of sediment-laden tributary into a lake can frequently be observed in detail. One can see the cloudy and thus denser tributary suddenly disappearing downwards, sometimes under strong foaming action, under the clearer water of the lake. Along the clearly defined line dividing the clear from the clouded water, flotsam frequently accumulates. The further path of the submerged density current along the lake bottom cannot be observed directly. In principle, the density current will disperse quickly when the bottom gradient approaches zero or even becomes negative or when the ratio ϱ/ϱ_w drops to about 1.0 due to sedimentation of the suspended load carried along.

On the other hand, the intensity of a density current can increase when it picks up previously deposited sediment and thereby raises its density. In reservoirs, through the continuous feed-back of this process, an avalanche-like increasing erosional process can get under way. According to Hartung (1959), velocities of up to $1.2 \, \mathrm{m \, s^{-1}}$ have been measured, which make the incorporation of previously deposited but still unconsolidated sediment quite feasible.

The behavior of saline solutions is similar to that of suspension currents. They can be observed in river mouths where under the influence of the tides the heavier sea water periodically intrudes below the fresh water. Save for strong turbulences along the boundary, a pronounced separation of the two media will generally develop, the salt water front attaining positions and shapes changing with the influence of the tides. This results in a two-dimensional current field in which the lower, saline layer is held in place through the friction exerted by the fresh water flowing off over it (Fig. 3.11).

As the salt tongue is assumed to be stationary ($v_2 = 0$) it acts like a weir with a wide top, above which the fresh water continues seaward. Plate (1974) had concluded from these consideration that $h_{2\max}$ is the height at which the fresh water flows at the critical velocity in relation to the salt water, i.e., $Fr'_0 = 1$. The critical depth then is

$$h_{1c} = \sqrt[3]{\frac{Q^2}{g'}} \ . \tag{25}$$

Experiments have shown that Eq. (25) does not reflect exactly the natural situation, as it is based on simplified assumptions. The velocities actually are not constant and there will always be some mixing. The real thickness of the fresh water at the end of the salt tongue will be about $h_1(0) = 1.2 \, h_{1c}$.

Calculation methods for shape and length of the salt tongue were given by Plate. In a usually satisfactory approximation the shape of the tongue can be

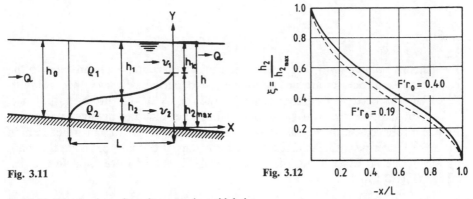

Fig. 3.11

Fig. 3.12

Fig. 3.11. Distribution of a salt tongue in a tidal river

Fig. 3.12. Shape of salt tongue. (From experiments of Keulegan, simplified after Plate 1974), no scale

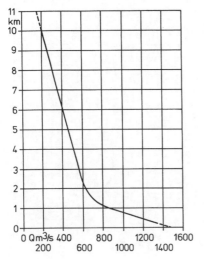

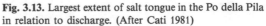

Fig. 3.13. Largest extent of salt tongue in the Po della Pila in relation to discharge. (After Cati 1981)

taken from a nomogram of Keulegan which was developed from experiments (Fig. 3.12).

The length L of the salt tongue can be calculated with the empirical formula

$$\frac{L}{h_0} = 1.06 \, (Re)^{1/4} (Fr_0')^{-11/4} \, , \tag{26}$$

with

$$Re = \frac{v_0 h_0}{v} \tag{27}$$

and

$$Fr'_0 = \frac{v_0}{\sqrt{g' h_0}} \ .$$
(28)

Given strong tidal currents, the river water can become completely mixed with the sea water. The processes are not completely understood. However, according to Plate they can be treated analytically with the theory of mixture by turbulence.

In principle also the configuration of the bodies of water involved has an influence on the length of the salt tongue. Sometimes "bags" of salt water remain stable for longer periods. As an example, the extent of the salt tongue in the main branch of the Po River delta, the Po della Pila, is shown in relation to the discharge (Fig. 3.13).

4 Sediments

4.1 Differentiation Between Bedload and Suspended Load

All matter contained in water – except for solute matter – is collectively referred to as solids. These consist of rock components of various origin, density, grain size, and shape. Floating matter of mostly organic derivation strictly speaking also belongs to the solids. However, quantitatively it is negligible compared to the mineral yield of the rivers and can thus be omitted from further considerations of river morphology, at least as far as central European conditions are concerned.

The solids are usually subdivided according to their mode of transportation. There is firstly the bedload moving downstream along the river bottom in rolling, skipping, or sliding fashion; and there is secondly the suspended material which, as the name indicates, floats along in suspension in the water. Distinguishing, for methodical reasons, between suspended load and bedload represents only a characterization of the actual state of movement, not a separation into two basically different types of behavior. Particles which under certain conditions may move along the bottom of the river and thus have bedload properties can be picked up with increasing turbulence and carried along in suspension until such time as they return to their original state of movement. If particles remain continuously in suspension, an equilibrium between the mean lifting forces due to turbulence and the uniform rate of descent of all particles in relation to the surrounding body of water has to establish itself.

As a condition for the transition between the two types of movement a limiting grain size is frequently introduced. Burz (1958) gives an empirical figure of 0.2 mm, Hartung (1959) assumes 1 mm, and Gallo and Rotundi (1965) suggest 0.35 mm. All these figures are, however, unsatisfactory, as they do not take into account the interaction between the various stages of movement. It was thus considered a step forward when the transition from bed load to suspended load transport began to be evaluated in relation to physical factors. According to Hayami (1941), the limiting condition for the relation between the settling velocity of a grain and the shear stress velocity (Sect. 3.1.1) is

$$\frac{\omega}{u^*} = \frac{\omega}{\sqrt{ghJ}} = 0.3 \ .$$
(29)

For the calculation of the settling velocity ω at small d ($= 0.1$ mm) the Stokes law can be applied:

$$\omega = \frac{1}{18} \, g d^2 \, \frac{\varrho_s - \varrho}{\eta} \ .$$
(30)

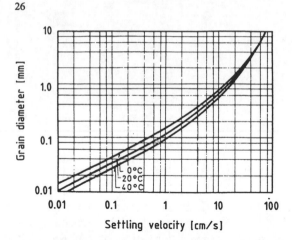

Fig. 4.1. Settling velocity of quartz in water plotted against diameter and temperature. (After Lane 1955)

It is, however, more convenient to use the diagram given in Fig. 4.1. When $\omega/u^* > 1$, generally only few particles will be found in suspension.

According to the investigations of Kresser (1964), it depends on the ratio between the impulse acting on the solids and their weight whether a particle will sink to the bottom or remain in suspension. If the shape coefficient and material constants are combined into one coefficient, this will be

$$K_1 = \frac{v_m^2}{gd} \ .$$

From calibration in various rivers Kresser found that $K_1 = 360$ will describe the boundary condition between suspension and bedload. From this the critical grain diameter is found as

$$d = \frac{v_m^2}{360g} \ . \tag{31}$$

His investigations showed further that the shape of the grains has only little influence on the magnitude of K_1.

Burz stated that Eq. (31) at $v_m > 1 \text{ m s}^{-1}$ will result in excessively large grain diameters. As shown by grain size analyses of suspended load samples taken during the July floods of 1965 in Bavarian rivers, the percentages of fractions above 0.2 mm − with few exceptions − are usually small. Burz concluded from this that the diameter $d = 0.2$ mm, recommended by him, in general correctly delineates the boundary between bedload and suspended load.

The differentiation between bedload and suspended load is a matter of methodology and should not tempt one to consider them as two clearly separated phenomena. It must not be overlooked that in natural rivers with fluctuating flow velocities, solids can change from the one type of movement into the other.

Some concluding thoughts on the question of the total transport of solids are given by Zeller (1963). He refers, for example, to Einstein, according to whose theory sediment cannot be held in suspension close to the river bottom when it is in an area where the eddies of the turbulent current become comparable in size

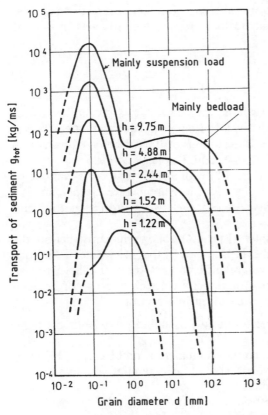

Fig. 4.2. Relation of total sediment transport of the Missouri and grain size and water depth. (After Einstein 1950)

to the respective grain diameters. The zone in question is called bottom layer. Without going into Einstein's mathematical formulas, reference is made to his example of the total sediment transport of the Missouri River (Fig. 4.2).

This clearly illustrates how the suspended load significantly increases during floods and that it can dominate the bedload by several orders of magnitude. The largest contribution comes from the size fraction around 0.1 mm diameter.

4.2 Bedload

4.2.1 Measurement of Bedload

The bedload budget and the formation of the river bed are closely interrelated. The best way to gain an insight into these natural formative processes, is to gather some impression of at least the bedload yield. In larger rivers direct measuring of the natural bedload transport is especially suitable for this purpose. From this, a so-called bedload function can be developed in relation to discharge or water level. When measuring the bedload transport, a number of limiting factors have to be taken into account, in contrast to suspended load measuring. This has led to different methods of determination. Amongst these factors are the grain sizes transported, the formation of bed forms, alternating transport directions in tidal areas, as well as the observation that bedload transport mostly sets in only at higher discharge.

4.2.1.1 Direct Bedload Measuring

Measurements with Retention Devices. One method that has proven itself over decades, is collecting the bedload at the river bottom in suitable containers (boxes, baskets) over a certain reference time. Descriptions of the devices used have been given by Smoltczyk (1955), Hubbell (1964), and Tippner (1981).

The Mühlhofer bedload sampler (Fig. 4.3) with a solid base is an example of the basket type. Outwardly similar constructions by Ehrenberger, Nesper, and the Swiss Federal Authority have a flexible base of loosely woven iron rings conforming to the shape of the bed.

The Helley-Smith bedload sampler (Fig. 4.4) follows in construction the Dutch "system Arnhem (BTMA)", like the one used by the Swiss Federal Authority. They work according to the pressure difference method and have proven themselves on sandy to gravely beds. For the construction and use of bedload samplers the following aspects have to be taken into account:

- in the front portion (inlet) the retention basket should exert neither a damming-up nor a suctional effect;
- it should fit closely to the bottom so that no material can be carried below it;
- the frame giving the necessary weight should be streamlined in shape, so as not to influence the retention process;
- during the measuring period the basket must not move from its position;
- it must be able to retain the maximum grain size expected.

SEDIMENT MEASURING DEVICES

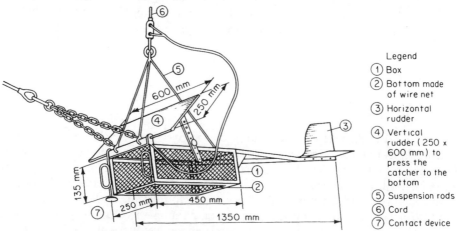

Fig. 4.3. Mühlhofer sampler. (Jarocki 1963)

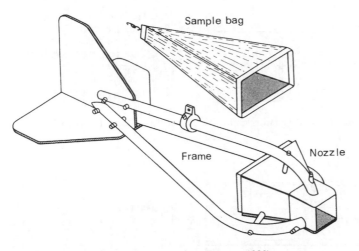

Fig. 4.4. Helley-Smith bedload sampler. (Emmett 1980)

The Arnhem-type devices are highly versatile and can be adapted to the specific situation encountered. Other more recent devices are the Karolyi and the VUV sampler.

A bedload pump which conveys the material caught in a funnel set in the river bed to an observation boat has been presented by Dornbusch (1965) as the Neisse sampling apparatus.

Bedload samplers can also be installed permanently at the river bottom, so-called traps or slots which collect the bedload and have to be emptied at certain intervals. Another method applied on the rivers Inn, Isar, Danube, and Salzach may also be mentioned here. Observations in consecutive river power plants tell us that the bedload will be sedimented upon the river entering the topmost reservoir and has to be removed there regularly. The mean annual bedload discharge can then be calculated from long-term records of the amount of gravel dredged.

Measuring by sampling devices is carried out at several positions across the section of the river similar to river gaging, so that the complete bedload-transporting width is represented by a number of sampling points. Other factors preferably determined at the same time are flow velocity and, if possible, also the suspended load at several points. In larger rivers such as the Rhine this can be carried out only from boats, whereas on smaller streams, measuring bridges or overhead cable crane installations for anchoring and operating have to be made available (Fig. 4.5).

The efficiency of a basket sampler depends on its specific design and for the devices mentioned above this is in the range of 70–90%, provided that the degree of filling remains low and no losses due to nonfulfillment of the above conditions are encountered. Measuring therefore has to be carried out with the utmost care.

As an example, bedload measurements in the Rhine at Karlsruhe-Maxau on October 10th, 1968 after Hinrich (1972) are given in Fig. 4.6.

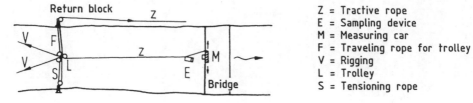

Z = Tractive rope
E = Sampling device
M = Measuring car
F = Traveling rope for trolley
V = Rigging
L = Trolley
S = Tensioning rope

Fig. 4.5. Bedload monitoring station Brienzwiler on the Aare River/Switzerland. (After Lichtenhahn 1977a)

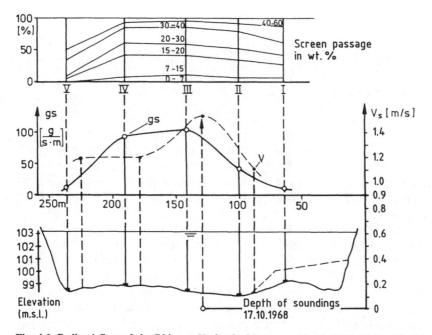

Fig. 4.6. Bedload flow of the Rhine at Karlsruhe-Maxau on 17. 10. 1968. (After Hinrich 1972)

Acoustic Method. The acoustic method is based either on measuring the magnitude of the sound of particle collisions during bedload transport with suitable underwater microphones, or of the noise caused by the bedload on contact with or impact on a metal body connected to a microphone. The method is based on Türk (1953) and has been perfected in recent years by the Federal German Bundesanstalt für Wasserbau in Karlsruhe. In the Rhine, for example, five acoustic receivers have been installed which register the impacts of bedload material in impulses per unit of time (Felkel and Störmer 1980). This is the only method allowing a continuous monitoring of the bedload transportation, although in a qualitative way only. However, the registration of the start and the end of bedload transportation itself is already important information. An interesting application was reported from the river Möll where a weir is lowered at the start of bedload transport (Schlatte 1979).

Underwater TV Cameras. Another method belonging to direct measuring or observation procedures is the use of underwater TV cameras. The bedload transport in the topmost bed layer which on gravel beds usually resists undisturbed sampling can be directly observed, recorded on data tape or photographed. The applicability is reduced during high concentrations of suspended load by the high turbidity of the water.

Tracer Methods. Addition of tracer pebbles. For an easier retrieval of the gravel transported in a control reach, foreign gravel of similar size and hardness may be used differing in color significantly from the usual bedload material of the river concerned. Already in 1929 tests were carried out in Bavaria by adding artificial pebbles in the form of blue concrete eggs of 4–5 cm diameter to the rivers Inn, Isar, and Lech. However, despite intensive searches, of the 45 000 pieces added to the Inn at river position 124.5 km, only 21 could be retrieved over a time span of 5 years and up to about 1 km downstream.

Radioactive Tracers. In this method natural or artificial sediments are used as an indicator material. Radioactive marking or activation according to Ajbulatov et al. (1961) can be carried out by:

– irradiation of sediments with neutrons in a reactor,
– adsorption of radionuclides on the surface of the sediment particle,
– inclusion of radionuclides in artificial material (glass), or
– incorporation of radionuclides in bedload particles.

The preparation, transport, and introduction into the river of the marked materials as well as measuring the resulting radioactivity require complicated and costly measuring installations in addition to careful radiation safety measures. This prevents the widespread use of this method.

Luminophoric Method. This method consists of spot-coating bedload or sediment particles with a luminescent pigment with the aid of a suitable cementing material and adding the material to the river when bedload transport begins. If a sample of bedload is then investigated under ultraviolet light, the marked grain will show up clearly and can be recorded. Tests with cementing materials such as agar-agar, bone glue, starch, gelatine, artificial resins, nitro-laquers, etc. have shown that these lead to agglomeration of individual grains. Water-glass resulted in the thinnest coating on the grains and the longest adherence. Rolling tests showed only minor attrition, so that the full luminescence was still preserved after about 160 km of simulated transport.

This is caused by the tendency of the pigment particles to become located in the surface pores of the carrier particles after repeated drying and washing during the application in the mixer. The colors yellow, red, green, blue, and orange were found to be suitable, so that it is possible to differentiate between materials added at different intervals. In tests along the Rhine marked gravels could be traced for up to 8 km below the point of introduction (Ruck 1979).

Magnetic Tracer Method. Recently there have been attempts to register the bedload transport in mountain rivers at a fixed point by induction loops with the aid of magnetic pebbles. For this purpose either naturally magnetic rocks are used which have to be investigated and characterized previously, or holes were drilled into pebbles and then filled with an iron core as tracer (Ergenzinger and Conrady 1982).

Measurement by tracers has become well established where the direction of transport is an important aspect, i.e., in tidal areas. For qualitative interpretations usually a large number of measurements are required (Mundschenk 1979). The advantage of the radioactive method lies in the direct mode of measurement, whereas with the use of the luminophoric method the samples have to be analyzed in the laboratory.

4.2.1.2 Indirect Bedload Measuring

River Bed Surveys. The oldest method of determining the bedload transport of a river is the topographic survey of deltas in natural and man-made lakes. It has to be remembered, however, that also the deposited suspended load has to be considered. The contribution of the suspended load can be determined from drill cores of the respective sediments. As an example, reference is made to a study of the delta of the Tyrolean Achen in the Chiemsee.

Based on repeated surveys and soundings, the growth of the delta can be traced with certainty back to 1869. Since then the average delta sedimentation is $140000 \, \text{m}^3 \, \text{a}^{-1}$, corresponding to an annual specific value of $148 \, \text{m}^3 \, \text{km}^{-2}$ for the catchment area measuring $950 \, \text{km}^2$ (Fig. 4.7).

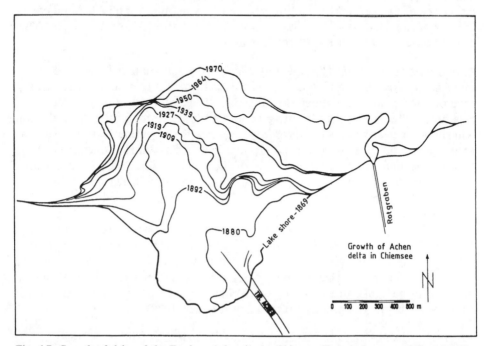

Fig. 4.7. Growth of delta of the Tyrolean Achen in the Chiemsee/Bavaria between 1869 and 1970

Table 4. Mean bedload transport of representative rivers ($m^2 \cdot a^{-1}$)

River	Locality	Drainage area	Period	Mean bedload transport	Reference
Rhine	Strassburg			250 – 300000	Gehrig 1977
Rhône	Donzère			500000	Gehrig 1977
Loire				800000	Gehrig 1977
Danube	Deutsch-Altenburg		1956/57	595000	Austrian survey
Inn	Rosenheim	10000	1961/76	102000	Dredging top of reservoir
Isar	Sylvenstein	1154	1958/83	53600	Silting survey
Saalach	Reichenhall	940	1960/79	100000	Dredging top of reservoir
Salzach	Haiming	6717	1953/77	122600	Dredging top of reservoir
Ammer	Ammersee	709	1948/55	25000	Bauer 1968
Tyrolean Achen	Chiemsee	952	1869/1964	40000	Delta survey

The delta was explored in 1965 – 68 at 41 locations over a rectangular grid by boreholes down to the basal lake clays in order to cover the complete body of detritus. From these data it became possible to trace the changes in the shape of the delta and at the same time to split the silting-up process into separate grain fractions.

The total growth in the 95 years from 1869 to 1965 was $13.2 \times 10^6 \, m^3$, consisting of 32% silt, 39% sand, and 29% gravel. It is assumed as an approximation that the gravel supplied to the delta roughly represents the bedload of the Tyrolean Achen, a mean annual supply of $(13.2 \times 10^6 \times 0.29) : 95$ or about $40000 \, m^3 \, a^{-1}$. Strictly speaking, also some of the sand should be considered as belonging to the bedload as it will not have been transported entirely in suspension. An exact subdivision is impossible, as fine-grained solids can be transported either as bedload or as suspended load, depending on the actual discharge conditions (cf. Sect. 4.1). Consequently, it is never possible to define exactly which proportion of the silting was solely due to the bedload. Nor must it be overlooked that the calculated mean value of $40000 \, m^3 \, a^{-1}$ is not a pristine value, as it has been affected by gravel dredging in the Achen and by river training. Table 4 gives some examples for the bedload discharge of rivers.

4.2.2 Sampling and Grain Size Analyses

In addition to determining the bedload transport of a river, the grain size distribution and the petrographic composition of the individual components are important research targets of river morphology. For a greater reliability of interpretation, sampling has to be carried out with utmost care.

Before sampling spots are fixed, it is necessary to know more about the factors controlling the bedload budget. Contributory factors to be taken into account especially are rates of bedload supply by tributaries, knick points of the bed, as well as any engineering-induced change of the bedload transport.

Bedload samples are commonly collected from dry gravel bars made up of younger deposits which show no signs of solidification. Mostly two to three samples are taken close to the shallow upstream side of the bar and on the side facing the water, in order to gain a representative picture of the bedload actually moved. Sampling the downstream part of the bar far from the water would distort the picture as at this position the fines are preferentially deposited during lower water levels. Gathering bedload samples directly out of the water would also lead to false results, as even with great care it cannot be avoided that during sampling fines are carried away by the current.

The surface of a stationary gravel bar only imperfectly reflects its shape during active bedload transport, its preserved morphology having been formed only during the drop in the water level. On its surface there is especially a lack of the fines carried away by the current, so that the remaining coarse fraction alone determines the morphology. Consequently, before sampling, the top 10–20 cm have to be removed. The actual bedload sample can then be taken from this uncovered interior of the gravel bar.

The sample thickness is usually 50–60 cm without penetrating into the zone moistened by groundwater. To be representative the sample must not be too small. For bedload of average coarseness usually about 100 l will be sufficient, whereas for coarser material twice this volume or even more will be necessary. The material excavated is stored in closed sacks. Although the interior section of the bar supplies the sample required for the grain size analyses, frequently also the cover layer is removed and screened for judging the degree of removal of fines and the degree of paving. Screening is rarely carried out on site, but mostly in the laboratory. The use of screening machines also for the coarse fraction now has become a matter of course.

The science of sedimentology has grown during the last few decades into a large and highly varied discipline. The strongest impulse was the search for oil and groundwater reserves. Only sedimentary formations are source and storage rocks for oil and gas, and as such they represent the most important aquifers. For questions of river morphology, however, only a few of the various aspects are of importance, i.e., data on particle size distributions, grain shape and composition, also rock-mechanical properties. Important are also a knowledge of the depositional environment and the resulting sedimentary structures.

There is a host of literature on performance, evaluation, and interpretation of grain size analyses. "Grain size is the basis of differentiating between clastic rocks. Grain size distributions can give insight into the transporting medium, as well as type and length of transport. In view of the highly irregular shape of the constituent particles, these data can be obtained by different methods – through determination of grain volume, weight, surface or cross-sectional area or through the settling velocity. Each method supplies only an approximation typical of the specific analyses" (Vossmerbäumer 1976).

The range of the grain sizes from the finest clay to coarse boulders extends over several orders of magnitude. This has to be taken into account when subdividing into size classes. The two most widely applied scales have to be men-

Table 5. Systems of grain size classification. (After Vossmerbäumer)

0,00001	0,0001	0,001	0,01	0,1	1	10	100 [mm]	

	Suspended matter		Pelite / Silt		Psammite / Sand		Rudite / Gravel	Boulder	
			Fine | Coarse		Fine | Coarse		Fine | Coarse		v. Engelhardt
Classification Designation	Clay			Silt	Sand / Medium sand		Gravel / Grit Medium gravel	Blocks	
	Fine clay		Coarse clay		Fine sand	Coarse sand	Fine gravel	Coarse gravel	
					0,063 0,63		6,3 63		
	0,0002	0,002	0,02		0,2 2,0		20 200 [mm]		

Clay		Silt		Sand		Gravel		Cobble	
		Fine | Medium | Coarse		Fine | Medium | Coarse		Fine | Medium | Coarse			DIN 4022

Diameter

0,02	0,2	0,63	2	6,3	20	63	200		[μ]
0,0002		0,002		0,02	0,063 0,2	0,63 2	6,3 20 63	200 [mm]	

Atterberg Scale: 3   2   1   0,5 0   -0,5 -1 -1,5 -2 [Z°]   Zeta-scale

Wentworth Scale: 10 9 8 7 6 5 4 3 2 1 0 -1 -2 -3 -4 -5 -6 -7 -8 [φ°]   Phi-scale

$$\frac{1}{1024} \quad \frac{1}{256} \quad \frac{1}{64} \quad \frac{1}{16} \quad \frac{1}{4} \quad 1 \quad 2 \ 4 \ 8 \ 16 \ 32 \ 64 \ 128 \ 256 \ [\text{mm}]$$

(Krumbein 1936 / See Walger 1964)

Clay		Silt		Sand		Gravel			Wentworth
Classification Designation				Very fine | Fine | Medium | Coarse | Very coarse | Granule		Pebble	Cobble	

tioned in some detail. In Germany, the Atterberg scale with a logarithmic spacing of, for example, $20-2-0.2$ mm is frequently used, whereas in the Anglo-American sphere the Wentworth scale is well established. Instead of the grain diameter in mm, in the latter scale φ values are used according to the relation

$$\varphi = -\frac{\log d}{\log 2} . \tag{32}$$

A transformation from the φ values to the Atterberg scale is possible. The respective notations for the latter are called ζ-values (Table 5).

For many different scientific and technical applications a number of additional screen scales have been developed which cover square- and round-hole screens with highly variable tolerances. A comprehensive overview of screen scales was given by Köster (1964).

Save for the raw tabulation, such data are evaluated in the form of graphic presentations: either as an area in a bar chart or in curves. Among the curves the

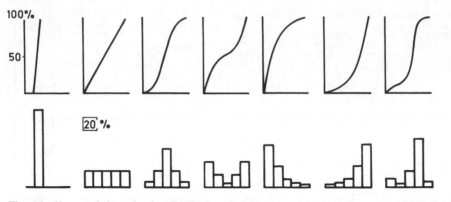

Fig. 4.8. Characteristic grain size distributions in histogram and cumulative curves. (After Marsal 1979)

cumulative curve clearly dominates the other methods of presentation such as grain distribution curve or frequency curve. The cumulative curve is frequently also called the grain size distribution curve, which, however, is not correct. For all presentations, depending on the purpose and the size range of the distributions, the abscissa is divided either normally or logarithmically. A double-logarithmic scale is rarely required for the grain size distributions encountered in river morphology. In contrast to this, the use of the log-probability plot can be of interest, albeit more for purely genetic questions than for normal grain size investigations.

The cumulative curve supplies most of the indicative values. On the ordinate it shows the accumulated weights of the various size intervals on the abscissa, i.e., the proportion of all particles having the same or a smaller diameter in the complete mixture. This proportion thus corresponds to the quantity passing through a screen of the respective mesh aperture or the sum of the grain size fractions in the screens below the particular screen. In the American literature frequent reference is made to "cumulative percent coarser than", i.e., the reverse of the above.

The indicative values are taken from the curve mostly through interpolation. Therefore in each case the exact points measured should be shown in the curve. This makes it possible to assign more weight to interpolated values closer to the exact values and thereby to improve the evaluation of a mixture (Köster 1964). In their cumulative curves highly mixed compositions differ notably from more homogeneous ones. The shape of the curve thus represents a characteristic which permits genetic comparison with other mixtures. Figure 4.8 shows a schematic comparison of characteristic distributions in bar charts with the corresponding cumulative curves.

The log-probability paper is preferred for the comparison of data in genetic interpretations. Its abscissa is logarithmically subdivided, the ordinate follows the bell-shaped Gauss curve. The cumulative curve of a log-normal distribution in this diagram will be a straight line. Such distributions are virtually unknown in nature. Consequently, more or less characteristic deviations from the straight line are developed. These are typical for certain types of deposits and of the physical processes involved. This results in indications for the distinction of eolian, limnic-fluvial, littoral, or truely marine sediments. A clear differentiation will not be possible with this method of grain size distribution alone. This needs a reasonable combination of petrographic and granulometric methods.

The probability paper used in the USA has a reversed subdivision of the abscissa, i.e., the smaller values start at the right. For comparison the respective sheets have to be viewed in mirror-image.

The grain size diagram frequently used in engineering (DIN 4190) is similar to the probability plot, but is only rarely applied in scientific investigations.

As mentioned already, a number of parameters can be taken from the cumulative curve for the comparison of various curves as well as for the better statistical characterization of a mixture. From the large number of possible arithmetic, geometric, and logarithmic values only the most commonly used will be mentioned here. A more comprehensive treatment is found, for example, in Köster (1964), Marsal (1967), and Batel (1971).

In addition to the mean diameter designating the arithmetic mean of a mixture or a fraction, the central value (median, middle value, most probable value, half-weight diameter, M or more frequently Md) is most commonly used. It is defined by that point on the cumulative curve at which 50% of the mixture belongs to a smaller and 50% to a larger value. The median permits a first comparison of various grain size distributions and their origin. It is directly controlled by the velocity of the agent which has deposited the sediment in question. According to Sindowski (1961), sands of various environments show typical values, as presented in Table 6.

Based on such considerations further subdivisions have been carried out. The size interval most prominently represented is often termed most frequent value or mode, while Bogolomov (1958) even calls the most prominent fraction mean grain size.

In German-language literature quartiles ("quarter weight diameter") are commonly used. The median divides the cumulative curve into two halves, the quartiles into quarters. They are defined by the grain size on the cumulative curve at 25 and 75% respectively. With the resulting three values, $Q_1 = 25\%$, $Q_2 = Md = 50\%$, and $Q_3 = 75\%$, various parameters such as sorting and skewness are defined (Table 9).

Table 6. Median grain size of various sands. (After Sindowski 1961)

Genetic type	Most frequent median (Md) phi-units	Most frequent range of median (Md) phi-units
Inland dunes	0.20 – 0.15	0.35 – 0.10
Coastal dunes	0.20 – 0.15	0.15 – 0.10
River sands	0.20 – 0.15	0.40 – 0.10
Tidal sands	0.15 – 0.10	0.20 – 0.05
Beach sands	0.20 – 0.15	0.25 – 0.10
Tidal channel sands	0.20 – 0.15	0.30 – 0.20

Table 7. Degrees of sorting

Sorting intervals (After Sindowski 1961)		Sorting intervals (After Füchtbauer 1959)		
S_o	Degree of sorting	Q_3/Q_1	Q_3/Q_1	Degree of sorting
Below 1.20	Very well sorted	Up to 1.5	Up to 1.23	Very good
1.20 – 1.50	Well sorted	Up to 2.0	Up to 1.41	Good
1.50 – 1.50	Moderately sorted	Up to 3.0	Up to 1.74	Moderate
Above 2.50	Poorly sorted	Up to 4.0	Up to 2.0	Poor
		Above 4.0	Above 2.0	Very poor

Table 8. Coefficient of sorting of various sands

Genetic type	Most frequent Sorting-value	Most frequent range
Inland dunes	1.30 – 1.20	1.50 – 1.30
Coastal dunes	1.30 – 1.20	1.40 – 1.30
River sands	1.40 – 1.30	1.40 – 1.20
Tidal sands	1.30 – 1.20	1.20 – 1.10
Beach sands	1.20 – 1.10	1.30 – 1.20
Tidal channel sands	1.30 – 1.20	1.20 – 1.00

Sorting indicates how many size fractions are present in a mixture. Steep cumulative curves incorporate only few size fractions, that are by definition well-sorted (Fig. 4.6). Flatter curves are made up of several size fractions and correspondingly considered as poorly sorted. The intervals for degrees of sorting as given in Table 7 have been found to be suitable. The coefficients of sorting of various sands as given by, for example, Sindowski (1961) and Lemcke et al. (1953) are compiled in Table 8.

The skewness or symmetry coefficient describes the symmetry or asymmetry in the sorting of the distribution. At a complete symmetry of the sorting, no skewness is developed, the respective parameter being zero, $S_K = 1$. Natural mixtures, however, nearly always show higher or lower degrees of asymmetry. The skewness is given as positive ($S_K < 1$) when the mean is more on the fine-grained side and as negative ($S_K > 1$) when it tends toward the coarser side (Figs. 4.9 and 4.8).

The parameters skewness and sorting consider only the range between 25 and 75% on the cumulative curve. In order to take into account also the fine and coarse portions, the so-called tails, sorting and skewness have also been defined differently (Table 9), and additionally the curtosis coefficient, K, has been introduced. In the original form given by Pettijohn (1957) it reads:

$$K = \frac{Q_3 - Q_1}{2(P_{90} - P_{10})} \ .$$

(33)

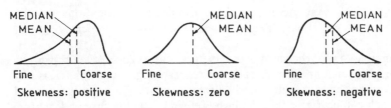

Fig. 4.9. Schematic presentation of skewness. (Vossmerbäumer 1977)

Table 9. Commonly used grain size parameters

Φ_{50} median (central value) (34)

Φ average (arithmetic mean)

$$\bar{\Phi} = \frac{\Sigma_v \Phi_v}{n}$$

Φ_m mean value after Folk (35)

$$\Phi_m = \frac{\Phi_{84} + \Phi_{50} + \Phi_{16}}{3}$$

σ standard deviation (36)

$$\sigma = \sqrt{\frac{\Sigma(\Phi_v - \bar{\Phi})^2}{n-1}}$$

δ inclusive graphic standard deviation after Folk (37)

$$\delta = \frac{\Phi_{84} - \Phi_{16}}{4} + \frac{\Phi_{95} - \Phi_5}{6.6}$$

S_o coefficient of sorting after Trask (38)

$$S_o = \sqrt{\frac{d_{75}}{d_{25}}}$$

S_k skewness coefficient after Trask (39)

$$S_k = \frac{d_{25} \cdot d_{75}}{M_d^2}$$

S'_k skewness coefficient after Folk (40)

$$S'_k = \frac{\Phi_{16} + \Phi_{84} - 2\Phi_{50}}{2(\Phi_{84} - \Phi_{16})} + \frac{\Phi_5 + \Phi_{95} - 2\Phi_{50}}{2(\Phi_{95} - \Phi_5)}$$

K curtosis after Folk (41)

$$K = \frac{\Phi_{95} - \Phi_5}{2.44(\Phi_{75} - \Phi_{25}}$$

P_{10} and P_{90} describe the abscissa values of the intersections of the cumulative curve with the 10 and the 90% line respectively, P stands for percentile, the Q values by comparison are then the 25th, 50th, and 75th percentiles.

Table 9 gives a selection of the most commonly used parameters derived from the cumulative curve, mainly in the convenient Φ-notation. The parameters of

Folk and Ward (1957) were developed from extensive investigations on the
bedload of rivers from gravel bars. The mean value nowadays is preferred against
the median.

It is also possible to plot the various parameters against each other, for example sorting against median, in order to gain further-reaching conclusions. It is furthermore well known that a mixture nearly always consists of several log-normal
populations which can be derived particularly well from the log-probability plots
(Walger 1962; Spencer 1963; Neumann 1963; Moss 1962, 1963). The turning
points divide the cumulative curve into several sections corresponding to the number of individual populations present in the mixture.

On the ordinate their proportion can be read off directly in percent. Walger
found in investigations of individual laminae as basic components of a sedimentary successsion that these also consist of three log-normal compounds which
could represent the three basic conditions of transport, i.e., rolling, saltation, and
suspension. Within the body of the sediment as the sum of the individual laminae
this threefold subdivision is encountered everywhere.

Spencer remarked that all clastic sediments consist of three log-normal grain
size populations, i.e., of gravel, sand, and clay with corresponding positions of the
medians. Their mixing is reflected in the cumulative curve and accounts for its
characteristic features. He confirmed the observations of Folk and Ward on recent
river sediments that they consist essentially of two portions, one coarse and one
fine, imparting a bimodality to the sediment. A third, intermediate, portion is
rarely developed, and if so, must always be considered as an indication of poor
sorting. Spencer talks about a grain/matrix relation. This bimodality is present
to different extents in all river gravels.

From combined shape and size analyses, Moss also observed three different
populations which he termed A, B, and C. They form a lattice work by mutual
support. Each arrangement was interpreted as representing certain hydrodynamic
conditions in combination with the petrographic peculiarities of the individual

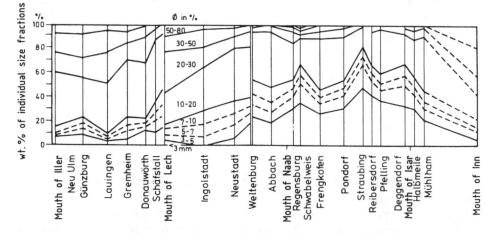

Fig. 4.10. Grain size graph of the Bavarian part of the Danube. (After Bauer 1965)

grains. He also observed mostly bimodal distributions. Such a distribution can lead to the lack of a certain size fraction in the cumulative curve and to two peaks in the bar chart. There are also a number of possibilities to assess the shape of the bedload grains, coarse as well as fine. The respective methods are rather time-consuming and laborious and thus inferior in execution to the shape estimation methods which are carried out more conveniently and rapidly. An experienced analyst will easily achieve the same accuracy in the statistical data.

For questions of river morphology investigations of the grain shape are rarely used. They are of more importance for scientific evaluations of the genesis of a given sediment. The derivation can in any case only be assessed by a petrographical analysis, as long as it does not follow directly from the outcrop situation (Sect. 4.2.3). A number of methods for shape analyses are given in the literature referred to. Of interest is also the development of the grain size distribution of the bedload, the so-called grain size graph, over longer reaches of a river. As an example the grain size graph of the Bavarian part of the Danube is shown in Fig. 4.10.

4.2.3 Derivation and Composition

The bedload of a river is evaluated according to different criteria by scientists and hydraulic engineering; for the morphologist the erosion is of main interest, for the engineer the mass per unit of volume. However, the question of the derivation of a bedload particle and thus its composition is a research target common to both fields.

The main erosion, according to climate and predominant lithology, takes place in the higher regions of the earth's surface. But also the main deposition of at least the coarser sediments takes place inland, for example in plains at the foot of the mountains eroded as long as these mountains do not directly end in the sea. Only a fraction of the eroded material reaches the seas immediately.

Erosion and accumulation alternate in space and time (Wundt 1953). Temporal alternation would be found at a given location where erosion takes place during floods and deposition during receding water levels (annual rhythmicity). More important is the spatial alternation through erosion from the hills into the valleys, from the mountain range to the plains below. This is particularly pronounced in arid climates, where physical erosion supplies vast amounts of detritus which can only be moved farther episodically during discharge resulting from sudden downpours. As furthermore vast areas in continental arid regions only show inland runoff, being cut off from the sea (cf. Chap. 6.1), huge amounts of sediments will accumulate in such continental depositories.

In contrast to this is the visualization of a stream based on the mechanics of transport, with erosion in the upper reaches, equilibrium in the middle, and deposition in the lower reaches. This simple model might be developed for short water courses, for example certain mountain torrents, but cannot be considered as the rule (cf. Chap. 5). The principle of alternating processes as an expression of the interaction of endogenic and exogenic forces over the whole length of the river is a better approach to the dynamic system "river".

There are three main sources of derivation of solids (cf. recent compilations of Bunza et al. 1976 and Lichtenhahn 1977):

1. Solid rocks. As solid are termed all those rocks which qualify as durable, this being controlled more by the behavior of the rock against the surface energy of the wetting water than by its hardness or resilience. Amongst the rocks which are virtually immune to attack by water are siliceous rocks, very pure limestones and dolomites, hard molasse sandstones, ortho- and paragneisses with low mica content, hornfelses, amphibolites, quartzites, granites, syenites, diorites, quartz-porphyries, etc. These are largely igneous rocks and silicified sediments, as well as highly metamorphosed varieties. Their most important properties are: dense texture, mainly homogeneous compact structure, and uniform size of mineral components. They are of constant volume, wall-forming, impervious to water, of low water absorption capacity, and weather mostly to large boulders.

2. Labile/solid rocks. These are susceptible to the surface energy of the wetting water. Transitions to durable solid rocks are possible due to their weathering to water-bearing decompositional minerals such as various micas, clay, kaolin, etc., or due to strong tectonic overprints.

Labile/solid rocks are marls, shales, argillaceous to calcareous marls, slates, all clay rocks such as those of the Jurassic and Cretaceous, rauhwackes (cellular dolomites), gypsiferous and saline rocks, mica schists and related rocks, phyllites as well as volcanic tuffs and ashes. Important properties are: porosity; bedded, schistous texture, splintery, crumbling, leafy, fibrous, scaly, predominantly water-bearing mineral components (swelling capacity), solidity based on labile-solid adhesion, low resistance to frost, poor load-bearing capacity, low to high water absorption with low water permeability, easily weathering.

3. Unconsolidated rocks. Amongst these are mainly boulder clays, terrace gravel, valley fills and reservoir material, debris cones, slope debris, and other debris bodies. Their properties are impermeability to high permeability to water, and highly variable grain size distribution.

The introduction of bedload material into streams usually occurs from several individual points along the banks (Karl 1970) rather than from longer stretches of erosion in the creeks themselves. For the river as a general water course the following possibilities exist for the derivation of the solids:

1. From the bedload supply of the tributaries which can be considerable (bedload injection, lateral debris cones).
2. From material eroded in bends (concave banks, general lateral erosion).
3. From downward erosion because of excess erosion potential.
4. From human influences (dumps without erosion protection, dam material, other measures such as river training, etc.).

The individual sources for streams are recorded cartographically within the framework of the morphological investigations (cf. Chap. 6.3.2.1). Given the knowledge of the geological situation of the catchment area and of the mechanical properties of its rocks, the bedload budget of a stream can be prepared. For the characterization of complete rivers or longer sections, the

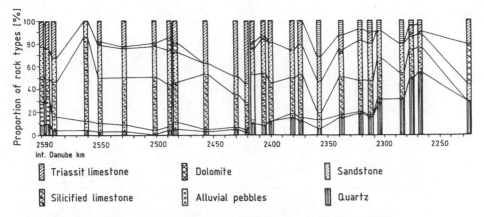

Fig. 4.11. Bedload graph of the Bavarian part of the Danube. (After Bauer 1965)

bedload samples taken for the determination of the transport volume and the grain size composition (Sect. 4.2.2) have to suffice.

The petrographic composition of the bedload does not entirely reflect the geological situation of a source area. This is easily understood, as the various rocks are of widely differing hardness and consequently the harder or more resilient ones will soon predominate over the softer ones, which, being rapidly worn down, lead to the relative enrichment of the former. With the aid of bedload or rock graphs which are combined with the grain size graphs, it is generally possible to trace this development. The influence of a tributary on the main stream can be noted with sufficient accuracy with the aid of the bedload composition, naturally all the better the more the rock spectra of the two differ from each other. In the Danube, for example, the in-put of a tributary coming from the Alps is always clearly recognizable. The bedload of the Danube is especially influenced by the Iller near Ulm or by the Inn near Passau (Bauer 1965). This, however, permits only the recognition of the tributary as a whole and not of its various contributing source areas, as the waters will process the bedload according to composition and discharge volume (Fig. 4.11).

Like index boulders in Quarternary glacial research, in the analyses of bedload material index pebbles are useful, although hardness and the resulting potential relative enrichment have to be considered. Consequently it is customary in the analysis to group together types of rocks having similar properties. The bedload graphs possess the disadvantage that they have to be prepared separately for various size fractions. In the Alpine tributaries of the Danube the characteristic difference between the light-colored, hard limestones (Wetterstein limestone, Rhaetoliassic limestones), which weather into larger blocks, and the Main Dolomite, which decomposes into smaller fragments due to its particular structure, have to be taken into account. Downstream the latter tends to become progressively concentrated in the smaller-sized fractions. In the flysch belt along the northern margin of the Alps, the influence of the hard silicified limestones in contrast to the softer, easily decomposable flysch sandstones and marls is just as pronounced.

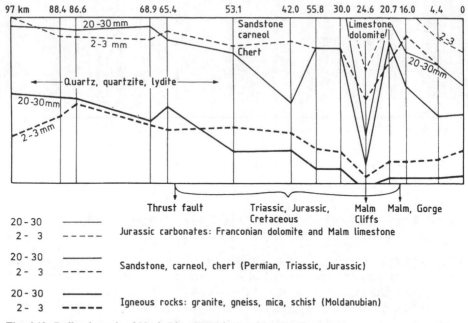

Fig. 4.12. Bedload graph of Naab River/Bavaria (total length of course)

The value of petrographic investigations will be illustrated by an example. For this purpose the Naab, a northern tributary of the Danube, was selected, as it carries material of quite different derivation and as at the same time it displays certain peculiarities. Figure 4.12 represents an attempt to illustrate in one diagram the overall situation based on data from 1964. It must be pointed out that the construction of various weirs and barrages considerably reduces the bedload transport, locally even causing a complete standstill, whereas the suspended load is largely transported through. The mixing chart is thus influenced by the artificially modified history of the river and has to be evaluated accordingly.

Based on geological and geomorphological criteria, the Naab can be subdivided into three sections:

1. Upper-Palatinate Forest with the Naab Hills and western foreland,
2. the slightly undulating foreland south of the Pfahl thrust fault with wide valley of the Naab,
3. the narrow valley of the Naab between Duggendorf and Etterzhausen.

The first two sections cannot be differentiated completely, as tributaries cross the Pfahl fault from both sides. The uppermost reaches, from the source to the Pfahl thrust fault approximately at river-km 65, are underlain by rocks of the Moldanubian. Its two source creeks, the Haide- and the Wald-Naab, come down from the southern slopes of the Fichtelgebirge. The Upper Palatinate Forest and the Naab Hills are made up almost entirely by igneous and metamorphic rocks such as granites, quartzitic gneisses, mica schists, quartz-phyllites, etc. The northwestern section, crossed by the Haide-Naab, is the Rotliegend-Triassic syncline of

Neustadt-Weiden with sandstones, carneol, and quartz. The latter is the predominant pebble material as it is present in ample supply in both areas as well as being a notable component in Pleistocene gravels.

In this first section the chart shows a rather uniform situation. The granite proportion varies between 34 and 50%, with coarser-grained varieties predominating. The sandstone content ranges from 4−14%, whereas quartz always accounts for more than 40% and locally even for considerably over 60%.

The second section extends from the Pfahl fault at km 65 to the entry of the river into the narrow valley section at km 20. The river here flows through a wide valley at the eastern margin of the Franconian Jura. The relief energy of this area is no longer large, the tributaries possess only low gradients and the supply of new bedload material is small. The granite proportion is reduced notably due to the lack of fresh supply. In the 20−30 mm fraction it drops from 50 to 9%, in the 2−3 mm fraction from 45 to 8%.

The absence of limestone or dolomite pebbles up to about km 30 near Burglengenfeld is noteworthy, as these rock types here are widespread. Calcareous material is introduced only at km 25, where it is derived in large quantities from a steep cliff bank consisting of Malm limestones. However, only a few kilometers downstream, below Kallmünz, the proportion of limestone pebbles has dropped already to 1%, locally even disappearing completely.

The lowermost section, from Duggendorf near km 20 to the mouth, is entirely incised into a limestone formation. There is an ample supply of material from the Malm limestones and the Franconian dolomite. The carbonate rocks here attain their maximum distribution in the 20−30 mm fraction with 41% and in the 2−3 mm fraction with 21%. This predominance of the softer sediments over quartz etc. can be explained by the fact that the material has been freshly introduced into the river which could not achieve a complete processing of it prior to reaching its connection with the Danube. This interpretation is supported by the low degree of rounding of the calcareous pebbles. At the mouth of the Naab granite and gneiss particles constitute only a small proportion. Micaceous and feldspar-rich varieties have disappeared completely, whereas quartz becomes enriched. Apparently here the granite fragments, especially the medium- to coarse-grained ones, had to take over the role of a softer rock in the presence of a large proportion of quartz.

One aspect unrelated to river morphology but affecting the composition of bedload and suspended load, as well as the chemistry of the river water, will be briefly referred to: the exploration for ore deposits. Due to the separation of specifically heavy ore particles from the lighter gangue, considerable enrichment processes can take place, especially when the long periods of time available are taken into account. A number of economic deposits, so-called placers, have been formed in this manner.

The ores of tin and monazite are almost exclusively extracted from placer deposits. The enormous Witwatersrand gold deposits of South Africa, for example, occur in highly silicified conglomerates resulting from fluviatile gravel. Similar concentrating processes can also take place in suitable marine bays such as in the case of the minette iron ores of Lorraine in France or the iron ores of Salzgitter north of the Harz mountains in West Germany. In poorly accessible areas the bedload and suspended load of streams as well as the chemical composition of the water are analyzed, and anomalies are traced upstream into the smallest tributaries until their source, hopefully an ore deposit, is located.

4.2.4 Attrition of Bedload Components

The bedload particles carried along by rivers are continuously reduced in size on their way, until in the lower reaches and deltas of the larger rivers they consist only of sand, silt, and clay. This process is called attrition. The size reduction of the bedload grains accompanies decreasing gradients. These connections have been known for a long time — Forchheimer (1930) refers to a treatise by Guglielmini (1665–1710) on the gradient of rivers — but it was Sternberg (1875) who eventually clad this observation in mathematical form. The so-called Sternberg's law implies that the attrition of a pebble is proportional to its weight in water and the length of the path x covered by it. In its modern version it reads

$$dP = cP \, dx \quad \text{or} \quad P = P_0 e^{-cx} \, , \tag{42}$$

c being a constant varying with the nature of the rock. If we assume that the relative shape of the rock during size reduction remains similar, the weight P and P_0 can be taken as proportional to the third power of the corresponding (idealized) diameters d and d_0, i.e.,

$$\frac{P}{P_0} = \frac{d^3}{d_0^3} \, ,$$

allowing a transformation of formula (40) into

$$d = d_0 e^{-ex/3} \, . \tag{43}$$

To find the coefficient of attribution, Eq. (40) is solved after

$$c = \frac{3}{0.434 \, x} \, (\log d_0 - \log d) \, . \tag{44}$$

As Sternberg pointed out, the bedload particle diameters can be determined only with difficulty and not with the required accuracy, as the shapes of the individual particles are more or less irregular. Consequently, only average values can be considered. When using these, however, the smaller grains have to be disregarded, as they are subject to different laws of movement than the bigger ones.

Sternberg's approach has been criticized repeatedly. According to Düll (1930), the process of attrition during movement of the bedload is rather varied. The fragments are subject to true attrition as well as to abrasion and destruction. Attrition in the strict sense is the loss of weight caused by friction with other pebbles during transport, whereas abrasion is the wear exerted on a stationary pebble by a pebble moving over it. According to this, large grains are more subject to abrasion than smaller ones. Breaking up takes place mainly in rocky narrows and steep sections of streams. But even without strong external forces, jointed rocks can disintegrate into smaller fragments. The resulting angular fragments then suffer from increased attrition until acquiring a rounded shape. Another destructive factor is atmospheric weathering. This weakens the structure of rock fragments, especially in the upper reaches of rivers where the gravel bars are more frequently exposed to alternate wetting and desiccation than in the middle and lower reaches.

It has to be kept in mind furthermore that the bedload material is rarely uniform in composition, being made up of materials from diverse petrographic sources. Their heterogeneous mechanical properties assist in sorting out the components. Because the softer particles are predominantly worn down, an apparent enrichment of the more resistant components takes place (cf. Sect. 4.2.3). In view of the manifold properties of the bedload in a river, Düll doubted whether it would be at all possible to trace this natural process of attrition with sufficient mathematical precision.

In order to facilitate a solution of the attrition problem, several authors, including Schoklitsch, have carried out experiments with rotating drums which contained bedload samples in a continuous flow of water. Düll described in detail set-up and results of such experiments at the Bavarian Landesstelle für Gewässerkunde. In order to correlate the results with naturally occurring attrition, some 16 t of basalt gravel of 40−50-mm grain size were fed into the Iller at Kellmünz. Samples retrieved later some 30 km downstream were compared with those that had covered the same distance in the rotating test drum. It was found that 2 km of transport in the drum corresponded to 1 km of transport in nature. From the large amount of data gathered during these experiments, a few examples will be given below (Table 10):

Table 10. Coefficient of attrition of various rock types. (After Düll)

Rock type	Coefficient of attrition c (per river-km)
Gray Triassic limestone	0.03 − 0.05
Treuchtlingen marble	0.06 − 0.08
Flysch limestone	0.03 − 0.06
Flysch sandstone	0.04 − 0.09
Light-colored dolomite	0.03 − 0.12
Quartzite with rough surface	0.02 − 0.04
Silicified limestones and sandstones	0.01 − 0.02
Quarternary Rhön basalts	0.02 − 0.05

Although care has to be taken when transferring the results of the drum experiments to natural rivers, they nevertheless facilitate an approximation of the order of magnitude of the coefficient of attrition.

Düll furthermore empirically developed a new law of attrition which includes the parameter n as a measure of the brittleness of the rock. His equation, however,

$$P = [P_0^{1-n} - (1-n)cx]^{1/1-n} ,$$ (45)

is not dimensionally correct and in the long run could not establish itself against Sternberg's approach, all the more as Schoklitsch revitalized these ideas and formulated strong support for them.

When Sternberg's law is applied to the solution of problems of river morphology, it is recommended basing the calculations always on several sample populations of the mean grain diameter. It must be ensured furthermore that the grain size graph of the sections investigated is not disturbed by tributaries with strong injection of bedload or other causes such as knick points at resistant rock ledges, etc.

The coefficient of attrition calculated with Eq. (44) does not consider only the pure attrition, but also all other factors which contribute to the size reduction of bedload particles, such as decomposition and shattering. There is some analogy to Strickler's velocity coefficient K_s (Chap. 3.1.1) which in addition to grain roughness contains a number of other functionally not exactly describable influences.

Disregarding the mathematical approaches, Scharf (1965) has compared the activity of the river with the wet-processing in a gravel plant and considered the river as:

1. A crusher. The coarse material is broken down into smaller pieces. This process takes place only in mountain torrents, waterfalls, and narrow gorges where coarse slope debris is obtained from steep cliffs into the river or its tributaries.
2. A ball mill. The action is by far the most important one as the bedload remains in motion. Slaty bedded material will be rapidly ground down just as all other soft rocks, or even material consisting of coarse feldspars such as coarse gneiss and amphibolite. They will not survive very long in a bedload population containing hard quartzose rock, limestone, and dolomite. These rock types can be considered as the active ball charge of a ball mill. They frequently show traces of impact. According to Scharf, dolomite is remarkably resistant. Although it is broken down rather rapidly into smaller fragments, these tend to stay angular even at small grain sizes, when they become enriched.
3. A mixer and classifier. The mixing action can best be observed below the entry point of bedload-bearing tributaries. The material introduced can remain stationary for some time or is only transported along the bank of entry and will become thoroughly mixed only after prolonged transport. At times the tributary even temporarily superimposes its own bedload spectrum onto the main stream. It is thus possible that there are differences in the bedload composition between the left and right banks of a river.

The classifying action results from the selection principle which can lead to relic gravels consisting of only the most resistant materials. It has been observed that the flat fragments of materials like mica schists, irrespective of their degree of rounding, can be transported more easily than round pebbles. This can lead to accumulation of mica schist pebbles in slow-flowing water and to the depletion of this component in reaches of rapid flow. In the latter section more rounded material will be deposited.

The imperfections of Sternberg's law led Stelczer (1968) to reassess the problem of bedload attrition. His conceptual model is based on the modern interpretation of the transport process (Sect. 4.2.6), according to which the particles migrate in surges along the bottom and are thus reduced in size in different manners. Attrition consequently consists of two phases of one process: the actual attrition of the moving particle and the abrasion on the stationary particle. Based on this situation the question arises as to over which period of time the particle covers a certain distance, and how much it is being reduced at the same time in size and weight. In order to obtain an answer to this, a new law of attrition has to be developed which must be substantiated by laboratory experiments and comparative field observations.

The factors controlling attrition were sudivided by Stelczer into five groups:

1. Mineralogic-petrographic composition of the particle. In contrast to Schoklitsch and Düll, who attempted to reach a solution through detailed investigations of the various rock types, Stelczer considers it as sufficient from an engineering point of view to subdivide the petrographic parameters into only four groups:
   - weathered volcanic rocks (andesite, andesitic tuff)
   - sediments (limestone, sandstone, marl)
   - metamorphic rocks (amphibolite, gneiss, mica schists)
   - quartz and quartzite.
   The rock types referred to in brackets have been investigated by him.
2. Frictional resistance. This is a function of the coefficient of friction, of the weight of the particle, and of a shape factor which expresses its degree of flattening. It is of great importance.
3. Impact force on collision. This will be situated somewhere between the extremes of the elastic and rigid impact and is basically the change of the impulse resulting from the mass of the particle and its velocity during the short duration of the contact (0.01 – 0.001's). The impact will be heaviest between a moving particle and one stationary on the bottom.
4. Flow of bedload. The probability of a collision between bedload particles grows with the volume of flow of the bedload.
5. Uniformity modulus of grain composition.

For the development of his equation, Stelczer used alternatively prismatic or spherical bedload particles. He furthermore distinguished between the attrition of moving particles and the abrasive action on stationary ones. In each case the mechanical work required by the specific force of friction is the governing factor. The presentation of the mathematical derivation is beyond the scope of this book. For the attrition of moving spherical bedload particles of the diameter d this leads to a formula which is similar to Sternberg's law as given in Eq. (43), with the only difference that the distance x covered is replaced by the traveling period or the period of lack of motion.

The attrition then is

$$d = d_0 e^{-k_1 t/3} , \tag{46}$$

and for the abrasive action on stationary grains

$$d = \sqrt[3]{d_0^3 - k_2 t'} . \tag{47}$$

The temporal sequence is shown schematically in Fig. 4.13. An average grain with the starting diameter d_0 moves during the time t, and is reduced in diameter by d_{m1} to d_1. It then stays at point 1 and does not move over the period t_1' during

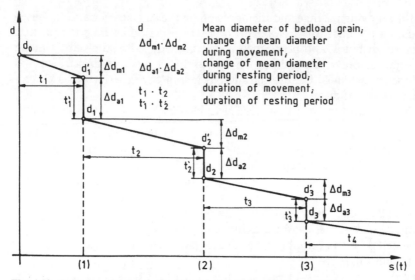

Fig. 4.13. Decrease of mean bedload grain size. (After Stelczer 1968)

which its diameter is abraded by d_{a1} to d_1. This is repeated during the subsequent intervals. For a random period it follows from Eqs. (46) and (47) that

$$\Delta di = d_{i-1} - \sqrt[3]{d_{i-1}^3\, e^{k_1^{1/3}} - k_2 t'} \ . \tag{48}$$

When several of the mostly rather short periods are considered, the values for d_i can be accumulated.

A number of laboratory experiments were carried out in order to give an indication of the attrition coefficients k_1 and k_2. To repeat Stelczer's detailed description of the experimental arrangement is not justifiable here; it suffices to state that he used a rotating drum, the inside of which was clad with a gravel layer embedded in concrete. The coefficient of attrition k_1 was calculated from the weight loss of the gravel sample and the weight of the suspended matter carried out. k_2 was determined with the aid of detailed stereophotogrammetric surveys of the fresh and of the ground-off inner surface of the drum. The evaluation of the test results led Stelczer to conclude that k_2 can be considered as a function of the bedload flow. He compiled the following values from the individual results (Table 11):

Table 11. Experimentally determined coefficients k_1 and k_2. (After Stelczer)

Rock type	k_1 (h^{-1})	k_2 (mm^{3}h^{-1})
Weathered volcanic rocks	11.1×10^{-3}	$0.718\, g_s$
Sediments	6.75×10^{-3}	$0.436\, g_s$
Metamorphic rocks	1.53×10^{-3}	$0.099\, g_s$
Quartz and quartzite	0.58×10^{-3}	$0.037\, g_s$

The dimension of bedload discharge per unit width of channel g_s is $g\,cm^{-1}\,s^{-1}$. Despite the theoretically convincing calculation of Stelczer's law of attrition, its practical application results in difficulties, since the period t of the movement as well as the stationary period t' cannot be determined exactly.

Starting from the low reliability of the usual laws of attrition, Gölz and Tippner (1985) undertook to verify the decrease of the grain size of pebbles in the Upper Rhine by a combination of attrition experiments in drums; also by petrographic analyses, grain size determinations, and measurements of the actual bedload discharge. They found that after some time of running, the attrition losses of carbonate rocks were about three times those of the other materials, whereas the rate of loss per unit of distance in both cases remained constant. The observed size reduction was ascribed to a lesser degree to attrition, but more to the sorting during transport.

4.2.5 Critical Shear Stress and Velocity

With every movement of water, shear stresses are transferred to the sides of the river bed. At uniform stationary flow the shear stress acting on the bottom can be calculated rather easily with Eq. (1). This is based on the concept that a unit bottom area is overlain by a water body with the weight $\varrho g h_z$. Its component $\varrho g h J$ parallel to the bottom must be in equilibrium with the shear stress acting on the perimeter of the channel.

At nonuniform movement the shear stress can no longer be calculated via Eq. (1) due to the additional forces of acceleration encountered. In an approach to the solution of the problem, Vollmers and Pernecker (1967) investigated the equilibrium conditions of an accelerated element of water on an inclined bottom (Fig. 4.14).

From the effective forces

$$G\sin\alpha + \varrho h_2 v_2^2 + \frac{\varrho g h_2^2}{2} - \varrho h_1 v_1^2 - \frac{\varrho g h_1^2}{2} - \tau_0 l = 0$$

there follows after several steps of calculation, and neglecting the h-members of higher order,

$$\tau_0 = \varrho g h_m (J_s + J_w) - J_w \varrho v_m^2 \ . \tag{49}$$

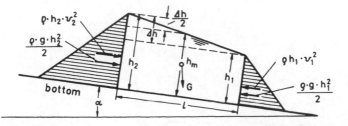

Fig. 4.14. Forces in an accelerated element of water. (After Vollmers and Pernecker 1967)

At retarded discharge the plus/minus signs of Eq. (49) have to be exchanged accordingly. The expression $J_w \varrho v_m^2$ is that portion of the impulse which reduces the shear stress at accelerated discharge and increases it at retarded discharge. It has to be remembered that J_w does not represent the absolute gradient of the water level but its inclination against the channel bottom. When applying Eq. (49) it is important to note that the bottom at times can change its position and thus its gradient, and that when the bedload discharge sets in, it may turn even opposite to the flow between certain points. Its true value in this case can be estimated from iterative calculations.

According to the theory of du Boys (1879, cf. Sect. 4.2.7) the beginning of bedload motion is clearly defined by the critical shear stress τ_c and/or the limiting velocity u_e resulting from it. Consequently bedload transport would set in when $\tau_0 = \tau_c$.

If this were generally applicable it would be of no concern for the start of the motion whether the shear stress τ_0 results from thin water cover and steep gradient or vice versa, as long as its numerical value remains unchanged. It is known, however, from tests and experience that the beginning of the bedload discharge does not depend only on the calculated shear stress on the walls but also on the relation between water depth and gradient. The theory of critical shear stress had to be re-evaluated. Shields (1936) presented a real break-through in showing that the critical shear stress depended on a dimensionless resistance number fc as a function of the Reynolds number. These relations are

$$f_c = \frac{\tau_c}{g d (\varrho_s - \varrho_w)} = f\left(\frac{u^* d}{\nu}\right) = f(Re^*) \tag{50}$$

and

$$\tau_c = F(Re^*)\, (\varrho_s - \varrho_w) g d . \tag{51}$$

The values for the function can be derived from Fig. 4.15. They are based on tests with uniform grains and consequently reflect the situation in rivers with mixed-grain bedload only with reservations. Above the curve, grains are in motion, below it they remain stationary.

Since Shields' paper a large number of data on the beginning of bedload transport have been published. An example would be Bonnefille (1963) on which the considerations of Vollmers and Pernecker (1967) are based. For the description of the critical condition three dimensionless parameters are used, i.e.,

the Reynolds number Re^* $\quad Re^* = \dfrac{u^* d}{\nu}$,

the Froude number Fr^* $\quad Fr^* = \dfrac{u^{*2}}{\varrho' g d}$,

and the "sedimentological diameter" d^* $\quad d^* = \left(\dfrac{\varrho' g}{\nu^2}\right)^{1/3} d$.

They form the relationship $d^* = \sqrt[3]{\dfrac{Re^{*2}}{Fr^*}}$.

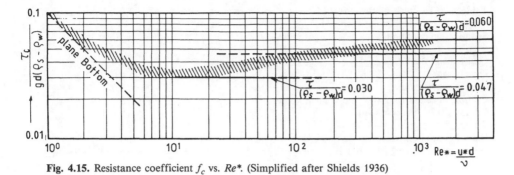

Fig. 4.15. Resistance coefficient f_c vs. Re^*. (Simplified after Shields 1936)

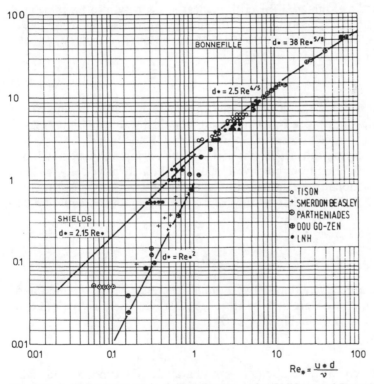

Fig. 4.16. Sediment transport in d^*/Re^* diagram. (After Vollmers and Pernecker 1967)

The relation between d^* and Re^* is shown in Fig. 4.16. Above $d^* = 2.5$ or respectively $Re^* = 1$, the beginning of sediment transport can be described by the straight line $d^* = 2.5\ Re^{*4/5}$ given by Bonnefille. From a transitional area at $d^* = 1$ onward, the straight line acquires the function $d^* = Re^{*2}$.

For very small sedimentological diameters the test results appear strongly influenced by the properties of the materials used. With decreasing d^* the func-

Table 12. Critical shear stress for different types of river bed layers. (After Wittmann)

Sediment composition	τ_c (Nm^{-2})
Quartz sand 0.2–0.4 mm	1.8–2.0
Quartz sand –2 mm	4.0
Coarse sand mixture	6.0–7.0
Compact bedded sand and fine-grained gravel	8.0–12
Loamy gravel (noncohesive)	15–20
Coarse quartz gravel 4–5 cm	44
Platy limestone fragments 1–2 cm thick, 4–6 cm long	50

tional dependence on Re^* becomes less well defined. This conclusion is supported by the differing position of the calculated Shields' function in Fig. 4.16.

For coarser bedload material the critical shear stress may also be calculated by placing the bedload transport in one of the functions discussed in Section 4.2.7 to zero. From the bedload function of Meyer-Peter in Eq. (67), for example, the following simple equation may be derived:

$$\tau_c = \beta d_m \ . \tag{52}$$

For a rapid estimation of the critical shear stresses in hydraulic engineering a number of empirical values are frequently used. For clear water the figures given in Table 12 may be applied according to Wittmann (1955).

Instead of employing the critical shear stress, the beginning of bedload motion can also be estimated from empirical value for the critical velocity, as both parameters are functionally connected with each other. From Eq. (2) for the shear stress velocity and Strickler's law of flow in Eq. (8) the following equation can be derived

$$v_c = k_s R^{1/6} \sqrt{\frac{\tau_c}{\varrho g}} \ . \tag{53}$$

There are furthermore several empirical formulas expressing v_c as a function of water depth and the mean grain diameter of the bottom, such as, for example, the equation of Schamov (in Tippner 1972)

$$v = 4.6 h^{1/6} d^{1/3} \ . \tag{54}$$

For everyday use there are also tabulations of v_c available (Wittmann 1955, Table 13).

The fact that the critical velocity is higher in sediment-laden than in clear water is surprising at first as one would expect higher shear stresses in sediment-laden waters of higher density. In reality, however, the shear stress decreases with increasing concentration of suspended matter. This can be explained by the situation that the flow velocity (cf. Chap. 3.1.2) at high suspended load drops much more rapidly near the bottom due to a higher viscosity than in clear water. Consequently turbulence is less pronounced at the bottom and the bedload grains are not as easily taken out of position.

Table 13. Critical velocity for different types of material. (After Wittmann)

Sediment composition	Size of grain	Critical velocity v_c (ms^{-1})	
		Clear water	Suspension-bearing water
Fine-grained sand	0.4 mm	0.15	0.25
Fine-grained sand	1.7 mm	0.35	0.45
Coarse-grained sand	2 - 3 mm	0.45 - 0.60	0.50 - 0.70
Fine-grained gravel	4 - 8 mm	0.60 - 0.75	0.70 - 0.90
Medium-grained gravel	8 - 16 mm	0.80 - 1.30	0.90 - 1.40
Coarse-grained gravel	16 - 25 mm	1.30 - 1.80	1.40 - 2.00

The shear stresses and flow velocities at which moving bedload comes to rest again, according to Wittmann are 10 - 20% higher than the values of Table 13. In older literature, for example Forchheimer (1930), the opposite is frequently stated. This is more plausible, because the initial force required to bring a stationary body into motion is greater than the force required to keep it in motion.

If the layer consists of cohesive material, the methods described so far for the determination of the critical shear stress no longer apply. It is well known, how-

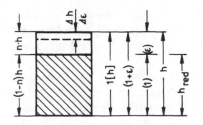

Fig. 4.17. Pore number and porosity of a soil element

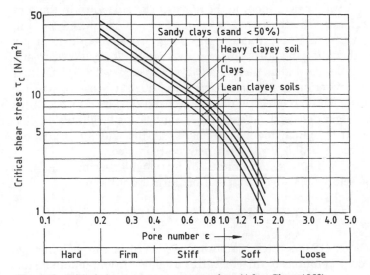

Fig. 4.18. Critical shear stress vs. pore number. (After Chow 1959)

ever, that the cohesive forces between colloidal particles increase with a decrease of the void ratio $\varepsilon = n/(1-n)$ with the porosity n (Fig. 4.17). The critical shear stress could thus be related to the pore number. Figure 4.18 from Ven Te Chow (1959) permits a rapid estimation of the order of magnitude of the critical shear stress for cohesive materials.

In contrast to river beds with sandy/gravelly layer, the critical shear stress cannot be used in the case of cohesive bottom material to obtain data on the deposition of suspended particles at decreasing shear stress. When particles have been lifted from the bottom, they largely assume the properties of suspended material and lose the cohesive forces which before had kept them within a more or less resistant bed.

4.2.6 Mechanics of Transport

Until quite recently the concepts of bedload movement were governed by the theory of du Buat (1816), according to which the bedload moves in layers at a velocity which decreases in a linear way downward. This is in contrast to the natural process as a gravel layer is never moved by the water as a homogeneous body, but individual grains are rather lifted from the bottom, moved a short stretch, and then moved again after a period of rest. As already observed by Schoklitsch, bedload movement takes place only in the uppermost layer of the bottom. Important insights into the mechanism of this transport were given by Welikanov (in Mayrhofer 1970), who was able to unravel the then still unknown process through careful observation.

The river bottom is usually not plane but rather shows undulations, especially in the case of fine-grained material. Any bodies above a conceptual reference plane are called bedforms. As shown in Fig. 4.19, a particle is eroded at the base of the shallower upstream side and moves rapidly over the entire upstream side. Once over the crest, it will be deposited on the downstream side to form the root zone of the next bedform. Whereas there is no motion of bedload directly in continuation of the downstream slope, it attains its maximum value at the crest. By internal relocation the whole bedform appears to migrate slowly downstream. Kinematic investigations show that there is always a system with the smallest possible velocity of migration established, and thus with the greatest possible height of the bedform. This is caused by the smaller, and thus quicker bedforms advancing onto the larger ones and causing these to grow to a height which cannot be exceeded under the given flow conditions. From the velocity of advance of the bedforms effective bedload transport may be estimated.

If g_s is the flow of solids across the crest, for a triangular shape (Fig. 4.19) Exner's relation of velocity of advance of the bedform would apply

$$u = \frac{g_s}{H} \, . \tag{55}$$

According to Führböter (1983), a well-established system of bedforms is formed in such a way that configurations of different sizes originate from random disturbances on the bottom, with the smaller bedforms advancing more rapidly than

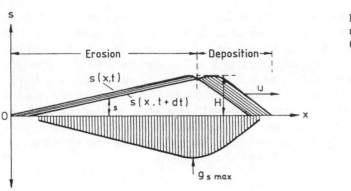

Fig. 4.19. Movement of material on bed forms. (After Mayrhofer 1970)

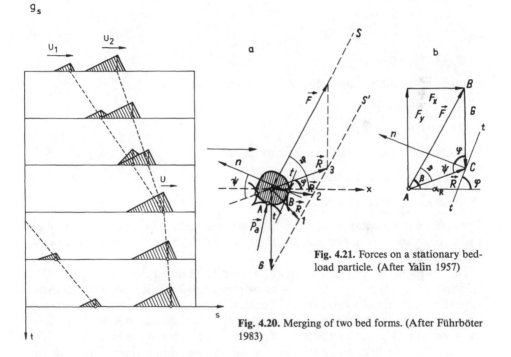

Fig. 4.21. Forces on a stationary bedload particle. (After Yalin 1957)

Fig. 4.20. Merging of two bed forms. (After Führböter 1983)

the larger ones, as is evident from Eq. (55). The smaller forms advance onto the bigger ones and merge with them (Fig. 4.20). This can be interpreted as a "synergetic process of self-organization". Taking into account the law of constancy of mass and Exner's relation, with the approximation g_s = constant, a so-called first-order unification operator can be formed, the repeated application of which leads to special functions. As shown by investigations of such unification factors, the results converge in such a way that after a certain number of time steps (in the experiments 500) from, for example, 1000 random starting disturbances each time a system of bedforms of nearly identical dimensions is formed.

A theoretical explanation of the interaction of forces on bedload grains was attempted by Yalin (1972). His approach will be briefly outlined with the aid of Fig. 4.21.

The bedload particle resting on a level bottom is influenced by the vectors of gravity G and a force F which is related to the velocity of the shear stress. The resulting vector R moves along the straight line S' parallel to S. If R passes between points A and B (position 1 in Fig. 4.21a) the grain rests with the force P_A in position A. If on the other hand the vector R comes to lie above point B (between positions 2 and 3), assuming cohesion-less conditions, A will not be loaded and only point B is subjected to a pressure which at times can lead to vibrations of the grain. It is removed from its position as soon as R forms the angle of friction ψ with the normal line n in position 3. According to the sine wave theorem it follows according to Fig. 4.21b with the vectors

$$\frac{F}{\sin{(\varphi+\psi)}} = \frac{G}{\sin{\beta}},$$

and   (56)

$$\beta = (\pi/2) - (\varphi + \psi - \vartheta)$$

as limiting conditions for incipient movement:

$$\frac{F}{G} \geq \frac{\sin{(\varphi+\psi)}}{\cos{[\vartheta-(\varphi+\psi)]}}.$$ (57)

In general the criteria for a moving grain are

$\varphi > \pi/2 - \psi$ for a saltation and

$\varphi < \pi/2 - \psi$ for rolling.

The angle ϑ is a function of the Reynolds number and of the velocity of the shear stress. The various relationships, which go beyond the present scope, were discussed in detail by Yalin.

In reality not every grain will leave its position immediately when Eq. (57) is fulfilled, as it may be wedged in by its neighbors and might have to be loosened first. This is especially the case for bedload without a sufficient proportion of fines. It can be illustrated by the statement that during fully developed bedload transport the finer components act as "ball bearings" for the larger ones. This also explains the observation that gravel bars in rivers from which part of the active bedload discharge is extracted by dredging, the resulting enrichment of the coarse components leads to a solidification of the remaining material which will be difficult to bring into motion again without outside "help".

After this outline of the forces acting on the individual grains we return to the problem of bed-forms. In addition to the well-known gravel bars in rivers with coarse bedload, there are the phenomena of ripples, dunes, and antidunes on sandy bed material. When discussing these, we must distinguish between subcritical ($Fr < 1$) and supercritical flow ($Fr > 1$).

1. Subcritical flow. As soon as the boundary velocity (Sect. 4.2.5) is reached, some grains start to roll until the whole surface is in motion. The bottom at first remains level. When the velocity increases further, sliding and saltation will set in, leading to the asymmetrical longitudinal section of the ripples

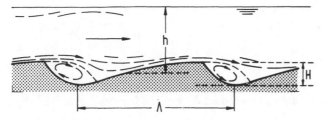

Fig. 4.22. Longitudinal sections of dunes and ripples. (After Yalin 1972)

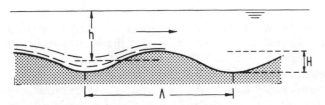

Fig. 4.23. Longitudinal section of antidunes. (After Yalin 1972)

(Fig. 4.22), sometimes with box-shaped surface structures. With further increasing velocity the ripples are replaced by dunes with a similar cross-section but longitudinal extent. In contrast to the ripples, the geometry of which cannot be mathematically expressed, wave length and height of the dunes are functions of water depth and flow velocity. With their flat upstream and steep downstream sides, the dunes are subject to formative processes similar to the wind-controlled coastal and desert dunes. They always advance in the flow direction of the water.

2. Supercritical flow. Under shooting or rapid motion, which in rivers is rare, antidunes with almost symmetrical cross-section are formed (Fig. 4.23). They can remain stationary or even migrate against the flow.

The physical causes for the formation of the characteristic shapes of the bedforms are still only incompletely understood. Their formation might possibly neutralize a certain surplus of the total transport capacity by continuous relocation of the bedload within the bedforms, i.e., despite intensive local movement, less bedload is transported downstream than would be the case on a level bottom. This interpretation is based on the minimum principle which is observed everywhere in nature.

An attempt to interpret the laws of formation of ripples and bars with the aid of the boundary layer theory was undertaken by Stehr (1975), who found that continuous transport of bedload along the bottom is possible only under constant shear stress on the bottom. This prerequisite, based on theoretical considerations, can only be fulfilled when the flow velocity continuously increases downstream, i.e., when the section of the channel passed through is reduced. This results in the at first absurd phenomenon that uniform transport of bedload along the bottom will take place only when the bottom rises in the direction of the flow. The rising upstream sides of the bedforms are thus nothing else than the expression of a constancy of the shear stress on the bottom. The flow velocity v_x in the external flow, i.e., outside the boundary layer, based on v_o under these conditions at $x = 0$ will be

$$v_x = v_0(x+1)^{0.075} \ . \tag{58a}$$

For the upstream side of a bar onto which flow is from one side only, the height s above the plane horizontal through the base point of the bar can be derived from the equation of Fig. 4.17 as

$$s = h_0 \frac{(x+1)^{0.075} - 1}{(x+1)^{0.075}} \ . \tag{58b}$$

In this, h_0 is the height of the accelerated layer of water. In shallow rivers it reaches the surface, i.e., h_0 is equal to the water depth h at the base point of the upstream slope. In deeper water $h_0 = 0.707 \, h$. The exact position of the boundary between these two possibilities cannot yet be decided.

After separation of the boundary layer behind the crest, on the upstream side a so-called natural slope forms below a horizontal eddy. Above a certain velocity the upslope traction exerted by the eddy no longer permits deposition on the downstream side: the grains fly over the eddy and the bar stops growing.

The geometry of bars in rivers has i.a. been investigated in detail by Allen (referred to in Nasner and Stehr). From field observations he found for the maximum height as a function of the water depth

$$H = s_{\max} = 0.086 \, h^{1.19} \ , \tag{58c}$$

whereas Stehr gave the equation

$$H = s_{\max} = 0.207 \, h \ . \tag{58d}$$

Equation (58d) appears to result in more acceptable heights.

A diagram developed by Yalin from numerous measurements is shown in Fig. 4.24. It should be noted that $Fr = 1$ defines the boundary between dunes and antidunes. For the uppermost portion of the curve ($Fr \geq 1$) Kennedy's equation (cited in Yalin) is valid

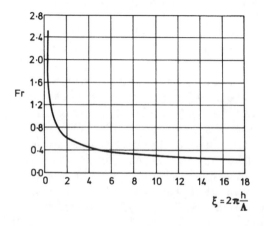

$$\xi = 2\pi \frac{h}{\Lambda}$$

Fig. 4.24. Fr vs. ξ. (Simplified after Yalin 1972)

$$Fr = \frac{2 + \xi \tanh \xi}{\xi^2 + 3 \xi \tanh \xi} \ ,$$

with   (59)

$$\xi = 2\pi \frac{h}{\Lambda} \ .$$

An alternative method for defining the limiting conditions for different types of bedforms was given by Bogárdi (1974). With the aid of the Froude number of the shear stress velocity

$$Fr^* = \frac{u^{*2}}{gd} \ ,$$

he found in laboratory experiments at 20 °C water temperature with model grains of 2.65 g cm^{-3} density a relation for the so-called stability factor, the reciprocal value of Fr^* which is

$$\frac{1}{Fr^*} = \frac{gd}{u^2_*} = \beta d^N \ . \tag{60}$$

As in these experiments $N = 0.882$ for each condition, the stability factor at a given grain size d is determined only by β. Solved for β one obtains

$$\beta = \frac{1}{Fr^* d^{0.882}} \ . \tag{61}$$

The following boundary conditions for β apply:

start of movement 550
sand ripples   322
dunes   66
transition   23.8
antidunes   9.65

For a further simplification of the application Eq. (60) can be solved for the shear stress velocity. It follows that

$$u^* = \left(\frac{g}{\beta}\right)^{1/2} d^{0.059} = \varepsilon d^{0.059} \ . \tag{62}$$

The relation between u^* and d is illustrated in Fig. 4.25. It is not completely clear whether the shapes of the bedforms observed in large rivers always fit into Bogárdi's scheme.

In tidal regions only smaller bedforms are regularly rebuilt by the tidal currents, whereas larger bedforms keep their morphology and present themselves as particular shapes typical of tide or ebb flow. The respective opposing current is apparently not strong enough or not of sufficiently long duration to destroy the previously established configuration.

Dillo (1960) described a section of the Elbe river in which on the one side the bars are orientated in the direction of the tide current, whereas on the other side they are orientated with the ebb flow. This is ascribed to the predominance of

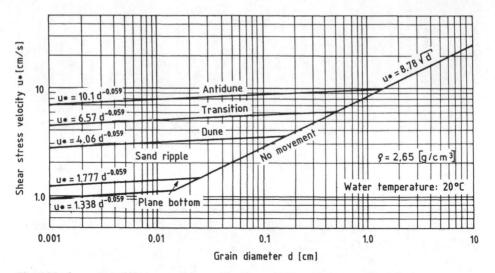

Fig. 4.25. d vs. $u*$ at different conditions of bedforms bodies. (After Bogardi 1974)

bedload transport during ebb and tide respectively on opposite sides of the river bed. The bars furthermore show considerable differences in size. According to the Wasser- und Schiffahrtsamt at Cuxhaven, the largest bedforms in the Elbe are up to 500 m long and 5 m high. Under such conditions Yalin's diagram (Fig. 4.24) no longer yields valid statements.

In regions of alternating tidal currents attempts have been carried out in recent years to determine the movement of sediment with the aid of radiometric methods. These allow direct measurements of even large-scale changes of the bottom morphology. For a quantification of the transport of solids it is necessary to know the thickness of the moving bottom layer and the transport velocity. The thickness of the layer in situ is determined with a scintillation probe according to the dispersive radiation method for which the vertical distribution of the radioactive tracer in the bottom can be considered as homogeneous or at a point only. The transport velocity, on the other hand is calculated by the movement of the center of gravity of the tracer "cloud" over a certain period of time.

During investigations of sediments in the estuary of the Elbe, Mundschenk (1979) could differentiate between two dominating transport mechanisms, a high-intensity boundary layer transport and a considerably less effective relocation of sediment throughout the complete moving bottom layer. Under normal hydrological conditions the boundary layer transport in the direction of the tide flow by far exceeds the movement of sediment with the ebb flow. Without clearly defining the boundary limits, Mundschenk gave its thickness as only 2−3 cm, whereas the moving bottom layer was estimated as up to 23.3 cm in depth.

4.2.7 Computation of Bedload Transport and Yield

du Boys' Equation. The first attempts at the calculation of the bedload transport
were carried out by du Boys (1879). He accepted du Buat's approach that transport proceeds by layers sliding over each other in a zone several grain diameters
thick. In each of the n layers of the thickness d the shear stress and the velocity
decrease downwards in a linear pattern (Fig. 4.26). Without taking buoyancy into
account, it is possible to calculate from this the bedload transport moved per unit
of time

$$g_s = \varrho_s g n d \, \frac{(n-1)\Delta \, v_s}{2} \, .$$

At the beginning of bedload transport the critical shear stress τ_c is equal to the
shear stress at the river bottom τ_0 and only the uppermost layer stays in motion.
In this case is $\tau_c = \tau_0$ for $n = 1$ and $\tau_c/\tau_0 = n$. Under these assumptions the equation for the bedload transport receives its usual form

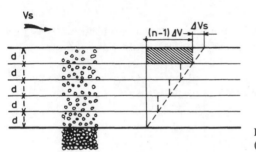

Fig. 4.26. Bedload transport after du Boys
(1879)

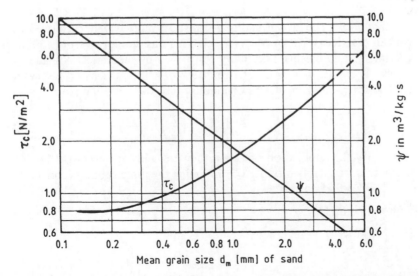

Fig. 4.27. ψ vs. τ_c for bedload transport equation of du Boys. (After Straub in Zeller 1963)

$$g_s = \psi \tau_0 (\tau_0 - \tau_c) \ ,$$ (63)

with the factor

$$\psi = \frac{\varrho_s g d \Delta v_s}{2\tau_c^2} \ .$$

Straub (in Zeller 1963) has investigated ψ and τ_c in a test flume channel for a number of practical cases (Fig. 4.27). Although du Boys' formula is based on incorrect physical premises, it has found wide acceptance and even today is sometimes still used for approximations of bedload transport.

Since then there has been no lack of methodical approaches to get away from the unsatisfactory basis of du Boys' equation, attempting to take into account the modern findings on the mechanism of bedload transport. A number of methods have been developed which, however, cannot be introduced here in detail. For an in-depth treatment of the relevant problems reference is made to the compilations of Bogárdi (1974), Graf (1971), and Yalin (1972). For a small selection the methods of Schoklitsch (1962), Meyer-Peter (1949), and Einstein (1950) shall be mentioned below.

Schoklitsch's Equation. From measurements in nature and laboratory test Schoklitsch found for the bedload flow per metre of width the equation

$$g_s = 2500 J^{3/2} (q - q_0) \ .$$ (64)

The boundary flow, at which the grain starts to move, is

$$q_0 = 0.26 \left(\frac{\varrho_s}{\varrho} - 1 \right)^{5/3} \frac{d^{3/2}}{J^{7/6}} \ .$$ (65)

As determining diameter Schoklitsch used d_{40}, i.e., 40% of the grains in a mixed sample are below this size.

Meyer-Peter's Equation. The law of bedload transport most widely employed in central European practice is that of Meyer-Peter. This was developed from 139 individual experiments and has been basically confirmed by theoretical considerations. This Swiss Formula, as it is usually referred to, will be discussed below in more detail. The form in which it was originally published in 1949 reads

$$\frac{\gamma R_s J_r}{(\gamma_s - \gamma) d_m} = \gamma \frac{Q_s}{Q} \left(\frac{k_s}{k_r} \right)^{3/2} \frac{hJ}{(\gamma_s - \gamma) d_m} = A'' + B'' \left(\frac{\gamma}{g} \right)^{1/3} \frac{g_s''^{2/3}}{(\gamma_s - \gamma) d_m} \ .$$ (66)

In order to simplify the application of this rather unwieldy formula, the law of bedload transport is frequently modified by inserting the dimensionless constants $A'' = 0.047$ and $B'' = 0.25$ to read:

$$g_s'' = 25.06 (\alpha h J - \beta d_m)^{3/2} \ ,$$ (67)

with $\alpha = \left(\dfrac{k_s}{k_r} \right)^{3/2} \dfrac{R_s}{h}$

and $\beta = 0.047 \gamma_s'' \ .$

As an explanation of the various components of Eqs. (66) and (67) it can be stated: The quotient k_s/k_r is the ratio between Strickler's bottom roughness k_s and the pure grain roughness k_r. At fully developed turbulence the latter is defined according to Eq. (9) as

$$k_r = \frac{26}{d_r^{1/6}} \, .$$

With the mean value found by Strickler for the calculation of k_s there is according to Eq. (10)

$$k_s = \frac{21.1}{d_r^{1/6}} \, ,$$

so that k_s/k_r for average conditions may be assumed as $21.1/26 = 0.81$. The quotient k_s/k_r is always below 1 and indicates that at shape roughness values of the moving bottom which appear in the form of ripples, dunes, or antidunes, part of the shear stress is not available for the transport of bedload.

Furthermore, the hydraulic radius R_s can be derived not only as usually from the cross-section of the channel, but also has to be put into relation to the partial flow Q_s which is effective for the bedload flow (Fig. 4.28). The broken boundary lines of F_s cut the lines of equal velocity at right angles.

As the position of the velocity contours is not exactly defined, it is recommended to use graphical aids for the calculation of R_s/h such as, e.g., the diagrams developed by the Swiss Amt für Straßen- und Flußbau given by Meyer-Peter and Lichtenhahn (1963). For trapezoidal sections it is possible to take

$$\frac{R_s}{h} \approx \frac{b_s}{U} \, . \tag{68}$$

The mean grain diameter d_m is calculated from the grain size distribution curves of the bottom mixture of the basal zone below the covering layer (cf. Sect. 4.2.2) according to formula

$$d_m = \frac{\Sigma p \Delta d}{100} \, , \tag{69}$$

where p denotes the proportion of screen passages at the relevant Δd (Fig. 4.29).

The applicability of the law of bedload flow to natural rivers with irregular beds is somewhat restricted. Mayrhofer (1964) comments: "In our Alpine rivers

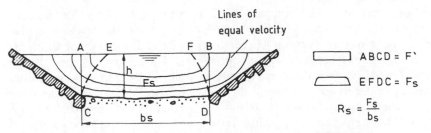

Fig. 4.28. Cross-section of geometry of river with parameters for calculation

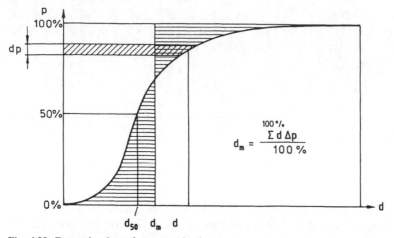

Fig. 4.29. Determination of mean grain size

with rather coarse-grained bedload for a small to average run-off, frequently gravel bars are observed not only stationary on the convex bank of bends, but also in straight sections migrating slowly and alternately on either side. In these sections the shape of the bed is characterized by pools below the banks opposite to the bars and by crossings with horizontal bottom and low water depth between the bars. At low water the flow runs from one pool to the next. Also at flood, when the current is more direct, the locally highly variable depths over pools and bars can be clearly noted. The bars migrate by erosion of bedload at the upstream and deposition at the downstream end. The bedload motion thus shows pronounced pulsations also in rivers with coarse material. In the area of the pools during low water the flow is concentrated in a narrow channel. It is here that there will be rather "vigorous" movement of bedload, even when the bars have not as yet been completely covered by water. With increasing water flow, strong movement of bedload starts on the bars despite the sparse cover, as locally steep gradients can develop. In the crossings the gradient at low water is rather high, but due to the great width of flow the specific discharge will be low.

A method had to be found which would permit the application of the law of bedload transport also under the conditions described. Meyer-Peter et al. (1935) recommended replacing the actual relief of the river bed by the so-called controlling cross profile. It is simply the hypsographic curve of the bottom of the sample section between the banks.

The horizontal smoothed bottom is used for a reference plane. For the construction of the hypsographic curves an arbitrary number of levels is placed parallel to the smoothed bottom. Their relevant widths are measured and averaged over all profiles recorded in the sample section. The controlling cross-profile is always of radial symmetry and can usually be made up from three straight lines (Fig. 4.30). In order to account for the differences in water depths as well as shear stress, the controlling cross-profile is split up vertically into trapezoidal strips (Fig. 4.31). In a strip with the width b during the reference time the volume of bedload transported will be

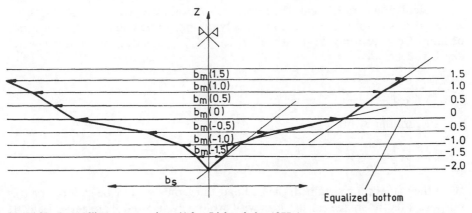

Fig. 4.30. Controlling cross-section. (After Lichtenhahn 1977a)

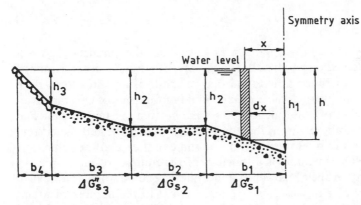

Fig. 4.31. Subdivision of cross-section according to strip method. (After Meyer-Peter and Lichtenhahn 1963)

$$\Delta G_s'' = \int_0^b g_s'' \, dx = 25.06 \int_0^b (\alpha \, h \, J - \beta \, d_m)^{3/2} \, dx \; . \tag{70}$$

With $\quad h = h_1 - \dfrac{h_1 - h_2}{b_1} x$

the integration results for a certain cross-section in

$$\Delta G''_{STr} = 10 b_1 \frac{(\alpha \, h_1 J - \beta \, d_m)^{5/2} - (\alpha \, h_2 J \beta \, d_m)^{5/2}}{\alpha \, (h_1 - h_2) J} \quad \text{for a trapezoid,} \tag{71a}$$

and

$$\Delta G''_{SRe} = 25.06 (\alpha \, h_2 J - \beta \, d_m)^{3/2} b_2 \quad \text{for a rectangle.} \tag{71b}$$

For the complete cross-section the individual values for all strips are accumulated:

$$G_s'' = \Sigma \Delta G_s'' \; .$$

This calculation is only valid when bedload motion takes place across the entire width. If this is not the case, the limiting width has to be determined first. Since bedload flow can only begin when $(\alpha h J - \beta d_m)$ is positive, the limiting water depth will be

$$h_{gr} = \frac{\beta d_m}{\alpha J} \, , \tag{72}$$

and the limiting width

$$b_{gr} = b_3 \frac{h_2 - h_{gr}}{h_2 - h_3} \, . \tag{73}$$

With the values thus obtained, the bedload function can now be set up. For this purpose a few reference calculations with arbitrarily selected h are carried out. If furthermore the influence of buoyancy is disregarded, the dry bedload flow will be

$$G_s = \Sigma \Delta G_s'' \frac{\varrho_s}{\varrho_s - \varrho} \, . \tag{74}$$

Einstein's Equation. Einstein's approach is based on the physically correct visualization that the bedload grains slide along on their way, or roll, or saltate. Based on numerous experiments, he found that the path of a grain with a given diameter can be split up into a number of individual movements of constant average length. The number of grains deposited on the bottom per unit of time and area depends on the grain diameter, the grain shape, its density, and the current conditions near the bottom. However, from the same unit area at the same time grains are removed. Their number is a function of the number of grains present and of the probability of their removal. Equilibrium of bedload flow is established when the number of particles thus arriving is equal to the number of grains being removed. Using the results of turbulence research Einstein developed for the transport of the grains the probability function

$$p = 1 - \frac{1}{\sqrt{\pi}} \int\limits_{-B^*\psi - 1/\eta_0}^{+B^*\psi - 1/\eta_0} e^{-t^2} dt = \frac{A^* \Phi}{1 + A^* \Phi} \, . \tag{75}$$

In this η_0 is the standard deviation of the so-called lift fluctuation or the buoyancy factor which in sediments of uniform grain size can be taken as 0.5. $A^* = 43.5$ and $B^* = 0.143$ are universal constants.

The rather involved Eq. (73) can be expressed by the functions of transport intensity

$$\Phi = \frac{g_s}{\varrho_s g} \left(\frac{\varrho}{\varrho_s - \varrho} \right)^{1/2} \left(\frac{1}{g d^3} \right)^{1/2} \tag{76}$$

and the intensity of movement

$$\Psi = \frac{g d}{u^{*2}} \frac{\varrho_s - \varrho}{\varrho} = \frac{1}{Fr^*} \cdot \frac{\varrho_s - \varrho}{\varrho} \, . \tag{77}$$

Their mutual relationship is illustrated in Fig. 4.32.

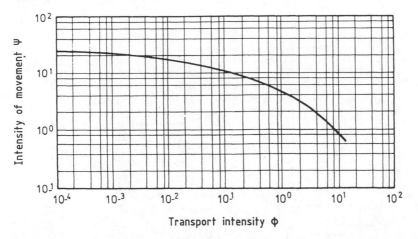

Fig. 4.32. Einstein-type equation of bedload transport relation. (Zeller 1969)

Equations (76) and (77) apply to sediments of uniform grain size, but can also be used for grain mixtures provided that certain substitutions not described here are carried out. In view of the conditions inherent in the calculation of the above-mentioned universal constants, Zeller (1963) is of the opinion that Einstein's bedload function is mainly suitable for rivers with fine- to medium-grained sediment and less for mountain rivers bearing coarse bedload.

From comparative considerations it was possible to transpose the Meyer-Peter equation into Einstein's presentation. According to Zeller it is

$$\Phi = \left(\frac{4}{\Psi} - 0.188\right)^{3/2} . \tag{78}$$

Within the range $10^{-4} \leqslant \Phi \leqslant 10^{1}$ both laws are in good agreement, provided bedload of uniform grain size is considered.

Like all other equations for the calculation of bedload transport, Einstein's law is also highly sensitive to variations of the controlling grain diameter. This is indicated by, e.g., the very shallow inclination of the upper branch of the Ψ/Φ curve. It is therefore advisable in every case to carry out several test calculations with different grain diameters.

Bedload Yield. As a last step, the calculation of the annual volume of bedload will be discussed. For this purpose the duration curve of discharge or water level and one of the above-mentioned bedload functions is used to arrive mathematically or graphically (Fig. 4.33) at a duration curve of bedload transport. In some cases it might suffice to use only the upper portion of the discharge duration curve, as bedload is usually transported only during a few weeks of the year.

The annual yield of bedload is presented by the area under the bedload transport duration curve in the first quadrant. During the period $t = 365$ days it will be

$$G_f = \int_0^t G_s dt . \tag{79}$$

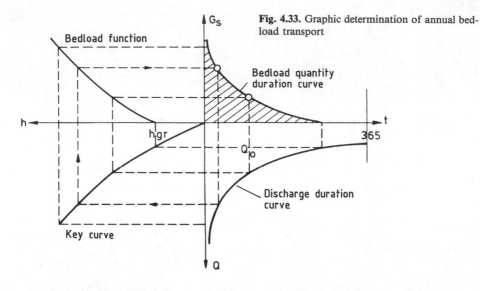

Fig. 4.33. Graphic determination of annual bedload transport

It is a misconception to assume that the problems of bedload transport have essentially been solved already. The uncertainty inherent in each calculation of the yield of bedload, despite the perfection of the more recent methods, has been expressed by Rouse (in Zeller): "Even if the equations for the bedload transport might be correct as such, it is difficult for the man in the field to apply them and to correctly collect the data required for his calculations. Depending on the investigator, in the same river section, differences of over 100% can be found in the calculation of g_s with the same equation for bedload transport." This rather discouraging statement, however, should not be taken as a reason for generally questioning the usefulness of bedload transport calculations.

4.3 Suspended Load

4.3.1 Measurement of Suspended Load

When at the turn of the century, the construction of hydropower plants spread in bedload-bearing rivers, more knowledge was needed about the suspended load for better estimating the silting-up processes likely to affect the respective reservoirs. Reference is made here to the *Guidelines for measuring suspended load* (DVWK Regulations for Hydraulic Engineering, 1986).

4.3.1.1 Individual Point Measurements

The distribution of suspended load in a channel is known from measurements to be rather variable and to undergo strong fluctuations. The particles suspended in

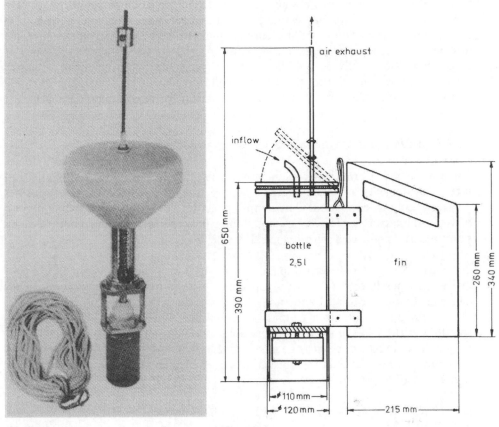

Fig. 4.34   **Fig. 4.35**

Fig. 4.34. Suspended sediment sampler. (Messrs. A. Ott, Kempten/FRG)

Fig. 4.35. Single suspended sediment sampler. (Bundesanstalt für Gewässerkunde, Coblenz/FRG)

the water frequently occur in a cloud-like concentration, so that two samples taken at the same location shortly after each other can differ considerably.

Because of the major technical effort required for continuous multi-point measurements, in Germany it was decided from the beginning to use a simple measuring method. Sampling is done with 10-l and 5-l buckets from points close to the surface and within the line of maximum velocity. The bucket is lowered to the water level by a rope from a bridge. By careful handling the water sample is drawn by raising the filled bucket without much spillage.

Instead of the sampling bucket, individual measurements may also employ the Ott sample scoop (Fig. 4.34). There is also a simple bottle device with inlet nozzle and air escape vent (Fig. 4.35) useful for the purpose.

In order to calculate the annual yield of the suspended load, a sufficient number of individual measurements has to be carried out. They are required specifically when changes in water level are anticipated. Given uniform water levels, up to several days may elapse between two measurements, but during the passage of the peak of a flood wave hourly measurements are necessary in order to cover all secondary minima and maxima. As the maximum concentration of suspended load usually occurs ahead of the flood peak, hourly measurements also have to be made during this period. Following Bavarian practice, as a rule of thumb, about 300 samples per year and station will yield reliable results.

4.3.1.2 Multi-Point Measurements (Full-Range Measurements)

For more detailed investigations and for verification of the individual near-surface measurements at certain times, and especially during floods, multi-point measurements are carried out. The sampling section is then split up into a number of vertical lines similar to river gaging along which at pre-determined depths one sample each is collected with a special device.

Ott Scoop Sampler. Greater sampling depths than with the simple bucket scoop method are possible with the Ott scoop sampler. This float-suspended device was designed by the Bavarian Landesamt für Wasserwirtschaft. A cylindrical galvanized float is screwed onto a 1-l or 2-l plastic bottle. With the aid of a clamp the sampling depth is set along a Bowden cable. The float remains on surface like a buoy whereas the scoop with the closed bottle is lowered into the water. When the clamp touches the float the bottle top is opened and then closed again by pulling the rope (Fig. 4.34).

Open-Tube Sampling Device. This device of the Bavarian Landesamt für Wasserwirtschaft (after Mücka) is based on the designs of Burz and Bauer and has been in use since 1968, its locking device having been improved considerably (Fig. 4.36). This construction replaces moving traps. There is instead a rubber hose drawn over both ends of the tube and secured with a hose connector. During sampling, the hose is pulled tight at both ends with a Perlon rope. This device is lowered on a measuring pole to the required depth. Because of its simple design it has proved useful for full-range measurements. Some problems occur over sandy stretches, and consequently a distance of at least 10 cm from the bottom has to be maintained. The rubber hose can be easily exchanged. Instructions for the preparation of this sampler can be obtained from the Bavarian Landesamt für Wasserwirtschaft.

Ott Suspension Sampler. Several years ago Ott (Kempten) introduced a sampling device for suspended load which accommodates six 2-l flasks and can be lowered vertically by a hoist to the required depth (Fig. 4.37). Behind the inlet nozzle there is a rotating inlet head which is electronically controlled to connect the nozzle with the next of the six flasks arranged in line in a box. Each flask is connected via a separate line to the distributing head. For directional stabilization a floating rudder is installed, while additional lateral correction is possible by lead weights attached to the sides. This device has the advantage that during full-range mea-

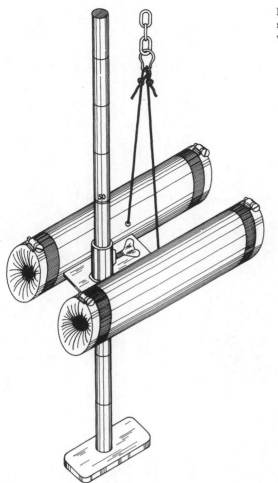

Fig. 4.36. Twin-pipe suspended sediment sampler. (Bayer. Landesamt für Wasserwirtschaft, Munich/FRG)

surements up to six samples can be collected in sequence without having to retrieve the sampler and exchange the flask after every submersion.

Pump Measurements. For multi-point measuring the water samples can also be taken by a pump device. The advantage of this is the simple combination of a sampling tube with hydrometric propellers at almost the same sampling point and the possibility of collecting samples of any volume desired (Christiansen 1974; Tippner 1981). This method is usually applied from survey craft working in larger rivers.

By pump adjustment or different diameters of the sampling pipes the suction velocity can be adjusted to the prevailing flow velocity.

Manning Suspended Solids Sampler. The portable automatic sampler model Manning S-4050 (Fig. 4.38) with its 24 plastic flasks each holding 0.5 l has been designed for taking individual samples. The device weighs about 16 kg and is

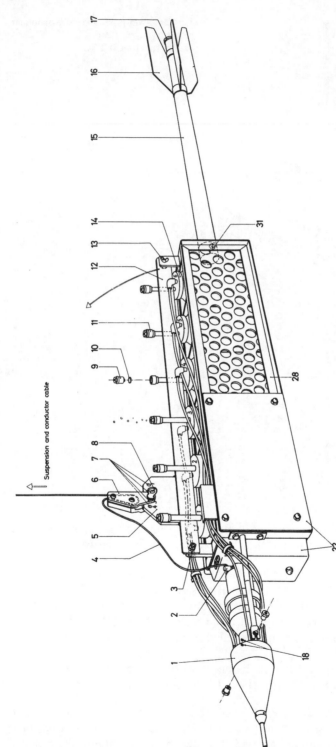

Suspension and conductor cable

Fig. 4.37. Suspended sediment sampler with six bottles. (A. Ott, Kempten/FRG)

Fig. 4.38. Portable automatic suspended solids sampler (Manning)

operated from a battery power pack. Control can be effected via clock, depth gauge contacts, water level switch, or by sensing the volume passing through a flow meter. The sample is drawn in by a vacuum pump and in order to avoid sedimentation in the inlet pipe, this is automatically rinsed before and after sampling. The Manning 6000 is a stationary model with 24 flasks holding 1 l each. A similar instrument (PB 10/T) is supplied by WTW Weilheim/Germany, employing 12 flasks of 1.8 l each.

4.3.1.3 Depth-Integrating Samplers

US Sampler D 49, Neyrpic Probe. For multi-point measurements the US sampler D49 may also be used (Fig. 4.39). In contrast to the Ott sampler, this is a streamlined device incorporating in its hollow space a flask of one pint or about 0.5 l. In the front there are three inlet nozzles with different openings which permit modification of the filling time. With this device integrating measurements are made, i.e., the flask is gradually filled while being lowered to the stream bed and raised back to the surface. To ensure a uniform rate of filling, a constant sinking and raising velocity has to be maintained. Because of the small sample volume and the rather cumbersome determination of the most suitable nozzle diameter, this instrument has not found wide acceptance. The Neyrpic probe (Fig. 4.40) which also delivers a depth-integrating sample, operates in much the same fashion.

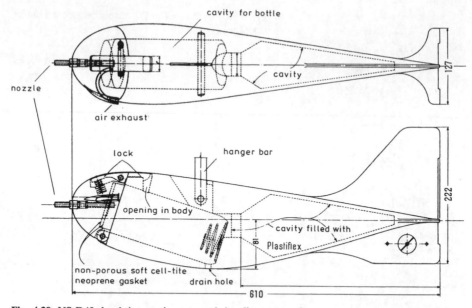

cavity for bottle

cavity

nozzle

air exhaust

127

lock

hanger bar

opening in body

cavity filled with
Plastiflex

non-porous soft cell-tite
neoprene gasket

drain hole

81

222

610

Fig. 4.39. US-D49 depth-integrating suspended sediment sampler

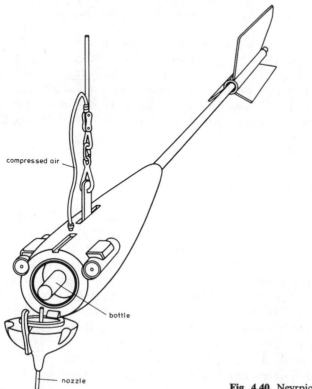

compressed air

bottle

nozzle

Fig. 4.40. Neyrpic sediment sampler

Filtering Bag Integrating Sampler. A very simple integrating measuring device for suspended load in smaller areas is the "filtering bag integrating sampler" (Fig. 4.41) described by Becchi et al. (1981). It works with an open wire basket which accepts a bag of tea filter paper (15×6 cm) with a welded-on plastic inlet nozzle, and can be mounted on a pole. The filter bag can accommodate up to 60 g of dry suspended material. Measuring times vary with the concentration of the suspended load and the discharge, and can range from a few hours to one week.

Cux-Sampler. The Cux sampler, developed by the Hamburg Behörde für Strom- und Hafenbau, is a scoop device which makes it possible to take samples from flowing and stationary waters (Fig. 4.42). It does not contain any electronic or mechanically moving parts, weighs about 200 kg and measures 3 m in length. The

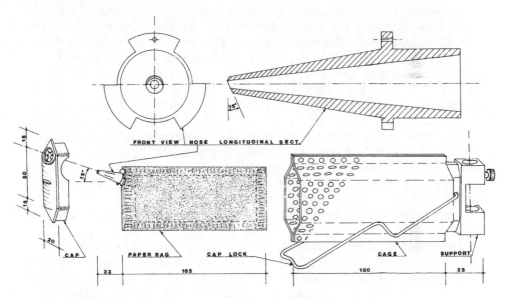

Fig. 4.41. Filtering bag integrating sampler. (Becchi et al. 1981)

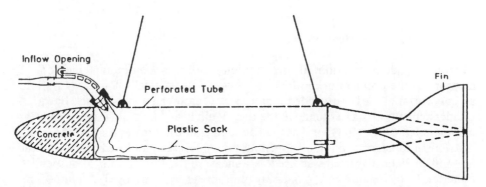

Fig. 4.42. Cux sampler (Strom und Hafenbau, Hamburg/FRG)

sample volume is 110 l. The sampler is produced for sample sizes of 5 – 150 l by Erich Berg, Hamburg.

The samples recovered are initially filtrated. To allow simultaneous processing of a larger number of samples, it is recommended to use filtration tables with 10 – 15 funnels with collecting buckets. As a support for the sensitive tops of the folded filters, each funnel should have a porous-entry sifting insert. The pore size of the filters has to be chosen so that on the one hand the vast majority of the suspended matter is retained, whereas on the other hand the process of filtration should not take excessively long. Medium- to wide-pored filter papers have been found to be practical. For filtration of the smaller samples from full-range measurements suction filtration devices are preferable to the funnels.

The suspended load of the water is calculated from the difference in weight between the dried used and the fresh filter papers and the volume of the respective water sample. For the determination of the annual yield, the product of suspended load and discharge has to be integrated over one year. This requires a considerable number of measurements which have to be carried out especially during major changes in water level and discharge. Whereas it is permissible during steady water levels to allow several days to elapse between measurements, during a flood period the intervals have to be reduced so that all important subminima and maxima of the flow trend are represented. In rapidly rising rivers hourly sampling might become necessary. In view of the fact that the suspended load is usually larger during the rise of the flood than at its peak and virtually always stays above that of the recession phase, measurements also have to be taken with sufficient frequency while the water level is dropping. For central European conditions a rule of thumb is that during a normal year about 300 samples allow meaningful results to be achieved (cf. Table 14).

For a conversion of the weight in which the suspended load is intically measured to volume units, usually an average density $\varrho_s = 1.35 \, \mathrm{g \, cm^{-3}}$ is used.

In order to speed up the handling of the suspended load sample and to obtain sufficient material for special investigations, stationary or mobile centrifuges were recently introduced. These instruments operate at up to 40 000 r.p.m. with cylinders of up to 2-l volume, and are suitable for a flow of up to 500 l h^{-1}. The separating efficiency depends on the volume of flow, which is adjustable according to the characteristics of the suspended matter (Hinrich 1965).

4.3.1.4 Continuous Measurements

Photometric determination of the turbidity offers a continuous method of measuring, the problem being the conversion of the optically found values into a concentration of suspended material. Measurements of the Bavarian Landesstelle für Gewässerkunde in the river Main near Marktbreit and in the Isar near Munich (Burz 1971) were not encouraging. In the former case a twin-beam instrument operating on the reflection principle was employed in which the amount of light dispersed was used as a measure for the amount of suspended material. In the second case, a one-way photometer was employed. An important source of error is the dispersion or diminution of the light beam which comes

Table 14. Annual denudation and suspended load of some drainage basins

River	Station	Size of upstream drainage area	Year	Average suspended load (g m^{-3})	(t km^{-2} a^{-1})	Reference
Elbe	Hitzacker	129877	1964/82	35	6.55	Jb.[a] Unteres Elbegeb. 1976
Weser	Intschede	37788	1970/83	35	12.6	Jb. Weser-Emsgeb. 1983
Ems	Versen	8469	1967/83	23	7.8	Jb. Weser-Emsgeb. 1983
Rhine	Kaub	103730	1970	53	37	Dekade Jb. 1970
Moselle	Cochem	27100	1970	23	44	Dekade Jb. 1970
Main	Kleinheubach	21505	1974/82	32	12.1	Jb. Rhein (Abschnitt Main) 1982
Neckar	Rockenau	12710	1972/79	42	29.9	Jb. SH Baden-Württ. 1979
Danube	Ingolstadt	20008	1973/82	37	19.6	Jb. Donau 1983
Danube	Vilshofen	47677	1967/83	22	11.8	Jb. Donau 1983
Naab	Heitzenhofen	5426	1973/82	27	8.0	Jb. Donau 1983
Inn	Oberaudorf	9712	1973/82	174	180	Jb. Donau 1983
Salzach	Burghausen	6649	1973/82	170	208	Jb. Donau 1983
Po	Casalmaggiore	42350	1928/41	–	346	Cati 1981
Secchia	Ponte Bacchello	1292	1924/41 60/64	–	1847	Cati 1981
Brahmaputra	–	580000	–	–	1370	Hadley and Walling 1984
Zambezi	–	1340000	–	–	75	Hadley and Walling 1984

[a] Jb. = Yearbook

with the grain size. At identical concentrations of suspended matter, a sample
with larger particles will allow more light to pass than one with finely divided particles. As the grain size distribution of the suspended matter in rivers is not constant, even calibration will not supply a useful relation between the turbidity measured and the concentration of suspended matter.

Long-term investigations of a river in hilly country (Nippes 1982) have shown
a good relationship between degree of turbidity and concentration of suspended
load if the evaluation differentiates between the various seasons and between rising and falling water.

Maintenance of the rather delicate instruments is another problem. There are,
however, already instruments on the market which require little maintenance such
as, e.g., devices which measure the turbidity in a free-falling jet of water.

The German Bundesanstalt für Gewässerkunde uses the dispersed-light probes
of Euro-Control. To exclude the influence of daylight the probe has to be installed
in an enclosed pipe (Reinemann et al. 1982).

The correlation between the concentration of suspended load and discharge
has been a subject of interest for some time. The attempts to find a function for
it were led by the desire to calculate the suspended load as easily from the discharge, as the latter can be derived via the discharge-rating curve from the water
levels. Usually an exponential function of the type

$$C_s = a Q^b \tag{80}$$

is applied, in which the parameters a and b are used to characterize the controlling
properties of the drainage area. To reduce the innumerable varieties of these properties to a few quantifiable parameters, Bogárdi (1956) used the following selection for the calculation of the exponent b:
average discharge MQ_1,
ratio between HQ and NQ_1,
size of drainage area A_E,
length of river from source to position measured.

Bogárdi (1956) obtained from these the required exponent b by multiple
regression correlations with the aid of the coaxial graphic presentation. The coefficient a cannot be derived from Eq. (80) on a physical basis, as it depends on
discharge and size of the drainage area. Bogárdi could show, however, that there
is a clear correlation between a and b, as well as between a and MQ_1, at least for
the Hungarian rivers, except the Danube and the Lafnitz. For rivers such as the
Danube, whose drainage area is made up of diverse subareas, such a correlation
is not possible.

Similar investigations of the suspended load in Bavarian rivers were carried
out by the Bavarian Landesstelle für Gewässerkunde (1972a). It was checked,
amongst other aspects, which of the following functions resulted in the best correlation:

$$C_s = a + b Q \tag{81a}$$

$$C_s = a e^{bQ} \tag{81b}$$

$$C_s = a Q^b \tag{81c}$$

Equation (81 c) corresponds to approach of Bogárdi. The results can be compiled as follows:

The correlation factors are usually low, indicating that the relation between suspended load and discharge is a rather loose one, and that an interdependence can be formulated mathematically only with great reservations. It is of little relevance here which of the three equations is used for the expression of the relationship. Apparently the factors controlling the suspended load are so diverse that simple mathematical expressions do not allow a satisfactory approximation. The double-logarithmic transformation [Eq. (81 c)] has usually given somewhat higher correlation factors, but this should not lead to the conclusion that an exponential function represents the most suitable equation for the relation. It has rather to be taken into account that the double-logarithmic transformation contains a systematic error, weighting the smaller values more than the larger ones. The range of the values for the flood intervals is thus apparently reduced, resulting in higher correlation coefficients.

An important source of error lies in the fact that the hydrographs of discharge and suspended load do not coincide, especially for the extreme values, but are rather shifted against each other irregularly. During the passage of the flood two

Table 15. Correlation between suspended load and discharge

Equation	Transformation	Regression function	Coefficient of correlation
(81 a)	Linear	$C_s = -43.7 + 13.16\ Q$	0.569
(81 b)	Semi-logarithmic	$\ln C_s = 3.38 + 0.079\ Q$	0.615
(81 c)	Double-logarithmic	$\ln C_s = 1.58 + 1.153\ \ln Q$	0.600

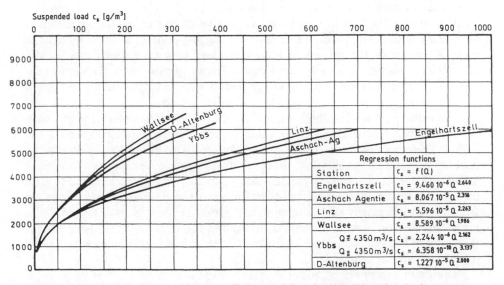

Fig. 4.43. Suspended solids in relation to discharge of Danube 1960/66. (After Gruber 1979)

identical discharge values can show completely different concentrations of suspended load.

As a first example the results of the regression analyses of the suspended load C_s (in g m^{-3}) of the river Ammer near Weilheim vs. the discharge Q (m^3 s^{-1}) are shown for the period 1950/66 in Table 15.

A second example illustrates the relationship between the suspended load concentration of the Danube and its discharge at several Austrian stations (Fig. 4.43) as found with the regression analysis based on Eq. (81 c). The diagram was presented by Gruber (1978) and was derived from measurements of the Austrian Bundesstrombauamt Vienna. It clearly illustrates the retaining effect of the barrages sited between the stations Engelhartszell and Deutsch-Altenburg.

However, prone these calculations may be to errors, the regression analyses cannot be dispensed with for approximative investigations, especially when it is necessary to obtain conclusions about rare events from a comparatively small number of data.

4.3.2 Grain Size Analyses

The investigation and evaluation of the grain size distribution of the suspended matter is basically not different from that of the bedload (Sect. 4.2.2). Screens are employed down to sizes of 0.063 mm = 4 ϕ, i.e., the accepted boundary between fine-grained sand and silt. Below these, sedimentation methods are used almost exclusively and at even smaller sizes, from about 0.0063 mm = 7 ϕ onward, centrifuges are employed. The value of 7 ϕ is already in the range between fine-grained silt and clay (boundary 9 ϕ, cf. Table 5). As an approximation it can be stated that the gravel and sand fractions are classified by screening, the silt fraction by sedimentation, and the clay fraction by centrifuging.

For investigations of suspended load usually the clay sizes are grouped together as one fraction in the cumulative curves or bar charts, in the latter as a rectangle with normal abscissa or as a calculated square in a semi-logarithmic scale. Screening of the suspended load at average flow velocities shows mostly fine-grained sand and silt, the clay fraction frequently not even accounting for 10% of the total weight. This can be ascribed to some extent to the sampling out of the water.

There are numerous methods for sedimentation, the principles of which have been described frequently, e.g., by Köhn (1929), Andreasen (1958), and Batel (1971). The buret method of Fabricius and Müller (1970) has proven itself as practical in execution and evaluation. It represents a further development of the pipet method (Köhn) and results in the same accuracy as the Atterberg method. With the aid of a computer program, all interesting grain size parameters can be derived from the automatically plotted cumulative curve. The program considers a size range from −6 ϕ (64 mm) to 9 ϕ (0.002 mm) and can thus also be applied for the screening of bedload samples. Usually the buret method is combined with screening of the coarser components.

The grain size analyses of bedload and suspended matter differ only in the method of separation of the individual fractions, which for finer-grained materials must necessarily be more refined. A size difference of 1 mm, for example, which could be insignificant for coarse cobbly glacial sediments or bedload, would be highly decisive for sand sizes.

4.3.3 Derivation and Composition

Rivers can only be enriched with bedload until their transport capacity is fully satisfied. For suspended matter the upper limit is given by the appearance of a paste-like viscous flow behavior. The suspended load is thus larger than the bedload (cf. Sect. 4.1). The ratio between these types of load varies from the mountain regions to the lowlands. In mountainous areas, depending on the source rocks, the respective proportions can be roughly equal, a situation also notably influenced by the discharge. Orders of magnitude of the concentrations of bedload to suspended load as given by Bauer and Burz (1968), Wagner (1960) and Lichtenhahn (1977a) are: Rhône at Leuk 1:1.2; Linth prior to entry into the Walen-See 1:3.5; Verdon (Durance) 1:3; Ammer river 1:1.4; Tyrolean Achen 1:1.3. In the Rhône entering Lake Geneva, the ratio is already 1:6.9 and at the mouth of the Alpine Rhine into Lake Constance about 1:30. Downstream the finer material rapidly starts to dominate. In lowland rivers only fine-grained gravel or sand are present which take over the role of the bedload (Bruk 1969; Schröder 1973). It is difficult to give exact data as the literature does not always clearly state where the boundary between suspended load and bedload has been drawn and no relation to the discharge is mentioned.

For the suspended load the source area weathering and the more widespread sheet-flows from the surface make themselves felt. For the bedload, individual, locally rather profusive point sources as well as more or less clearly recognizable reaches of erosion can be considered as supply sources. In contrast to this, for the suspended load in addition to the above sources the attrition of pebbles (Sect. 4.2.4) and the general sheet-flow from weathering products, soils, and other un-consolidated material have to be considered.

There are suspension-rich creeks and streams which are supplied from only one or a few point sources or even from only one tributary, and others which have as a source more or less the complete source area and thus represent a "collector" channel.

Such origins frequently find their expression in the names of rivers. Red River is the name given to several rivers, predominantly in arid areas, which obtain their coloration from red sandstones, marls, and lateritic soils. The best known example is the Colorado River which annually transports some 250 million t of sand and suspended matter into the Gulf of California. Similarly known is the Hwang-ho (Yellow River), which flows through vast areas underlain by loess, and which next to the Amazon and the Ganges-Brahmaputra system, is one of the most suspension-rich rivers of the world. The White Nile derives its milky turbidity from the Sobat. White in this context can also imply "clear" or "colorless" as well as the presence of a light-colored sandy bottom or light-colored, mostly calcareous bedload material. The name Weißbach, so widespread in the Alps, is a frequent indication for this material.

The situation is similar for brown- or black-water rivers, which frequently are rather clear and owe their color almost exclusively to organic substances. The best-known example of this is the Rio Negro, the large northern tributary of the Amazon. Vareschi (1963) reported on interesting relations between white- and black-water rivers of the Orinocco and Amazon regions. On the other hand, Rio Negro can also imply dark compounds derived from easily weathered rocks such as in the Rio Negro of Patagonia, carrying material from weathered basalts. In central Europe there are also numerous similar connotations such as, e.g., the Black Lütschine in the Berner Oberland, the Black Nolla in Graubünden, both derived from dark labile/solid Bündener Schists. Next to the Schwarzachs (Black A.), Weißachs (White A.), and Rot(t)achs (Red A.) there are names such as Kotach (Mire A.) or Trübenbach (Murky Creek) which refer to at least a passing heavy supply to suspended material. In contrast to this, the Black Regen in the Bavarian Forest, for example, is a brown-water river which is actually clear. The Red Main, one of the two source rivers of the Main, carries much sand and fines derived from Keuper sandstones, the Feuerletten (marly shales), Dogger sandstones and marls south of Bayreuth. The White Main, on the other hand, originates in the igneous rocks of the Fichtelgebirge mountains. Blautopf (Blue Pothole) is a clear karst spring near Blaubeuren, Dreckwalz (Dirt Roll) a creek in the loess area of the Kraichgau, both in southwestern Germany (Wagner 1960).

The regional petrographic situation is in parts much influenced by climatic factors, topographic elevation, and vegetational cover. There has been no lack of attempts to classify the corrosion, the general erosion (bedload, suspended load, and dissolved solids) according to climatic zones as the run-off according to diverse types of regimes (Chap. 6.3.4). The regional differences in origin of suspension load can be illustrated by the Po as an example (Fig. 4.44). Whereas its Alpine tributaries only rarely supply more than the equivalent of 0.1 mm a^{-1}, the erosion in the Appenines is locally more than 8 mm a^{-1}.

Not only the geographical differences in supply, but also the temporal fluctuation of the suspended load can be characteristic. The transport of suspended material is subject to pronounced seasonal variations, while in glacial creeks there are even diurnal changes. This is governed by the respective type of regime and in certain cases also by petrographic peculiarities. There are simple and complex regimes. The annual averages consequently have to be interpreted as statistical values only.

In contrast to the bedload, suspended load is transported throughout the year. Even clear water contains small amounts of suspended matter, which to the naked eye are barely perceptible or invisible. It is known from experience that a concentration of suspended load below $10-15 \text{ mg l}^{-1}$ can no longer be recognized as turbidity. The bulk of the transport, however, can be essentially concentrated over the few months of higher discharge, depending on the type of regime. In smaller rivers or creeks it may even be restricted to a few occurrences of floods. In larger collecting streams, which as complex types of regimes exert a more equalizing influence, and in rivers which flow through regions yielding much fine-grained sediment, the individual downpours or seasonal fluctuations are not so pronounced. Such rivers always display some turbidity.

Figure 4.45 shows the suspended load of the Bavarian Danube and some of its tributaries over a 25-year period (1930/55). There are clearly two types: (1) The

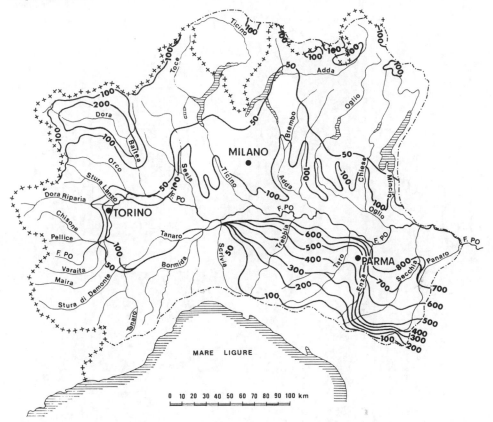

Fig. 4.44. Contour lines for mean annual removal of suspended solids in discharge basin of the Po in mm×10^{-3} a^{-1}. (After Cati 1981)

Danube down to Passau as the collector of the northern tributaries low in suspended load (Suevian-Franconian Jura, Bavarian and Upper-Palatinate Forest) as well as from the northern Calcareous Alps (Kalkalpen) and their forelands the rivers Iller, Lech, and Isar. In comparison to the Inn, these are not notably rich in suspended load. (2) The Inn as the main collector of a large part of the eastern Alps, imparting at Passau to the Danube its characteristic imprint with regard to the suspended load (cf. Chaps. 5.3.2 and 6.4). Its high suspended load exceeds that of the similarly structured Salzach and adds to its predominance over the Danube at Passau. It results from a variety of sources, making the Inn a real collector: glaciated regions of over 4000 m elevation in the source area (Bernina); wide, partly glaciated areas above 3000 m; tectonically highly overprinted crystalline regions (parts of the Silvretta, Verwall, Ötztal-Stubai crystalline core, Zillertal Alps, Hohe Tauern) and calcareous regions (Engadine Dolomites, northern Calcareous Alps); labile-solid Bündener Schists (Engadin Geological Window); numerous unconsolidated glacial sediments. Its tributaries, large rivers themselves, are all rich in bedload and suspended load, especially the Rosanna, Trisanna, Pitztaler Ache, Ötztaler Ache, Melach, Sill, Ziller, and finally the

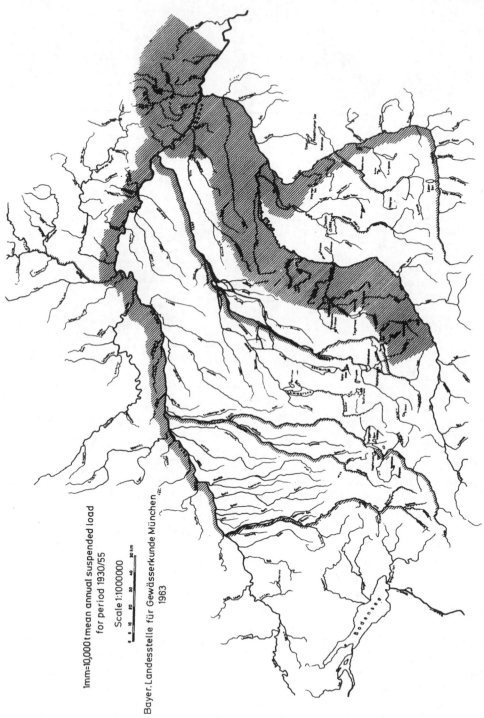

1mm=10,000 t mean annual suspended load
for period 1930/55

Scale 1:1000000

Bayer. Landesstelle für Gewässerkunde München
1963

Fig. 4.45. Suspended load content of Danube and its major tributaries

Salzach, as they are fed by numerous mountain torrents. This results in a mosaic of point sources and erosional reaches distributed along the rivers themselves.

The sediment-rich Alpine Rhine and Rhône shed their load almost completely in Lake Constance and Lake Geneva respectively. Functionally similar to the Danube are also the Drau, Piave, Adige, Isère, Durance which, save for the Piave and Adige, influence their main rivers Danube and Rhône (below Lake Geneva) correspondingly.

The suspended load charts of a river maintain their indicative value all along, as the suspended matter is retained only insignificantly by structures in the river unless the river passes through large reservoirs.

The question as to the source of the suspended load can be answered conclusively in that it is identical to that of the bedload, that it is formed by the attrition of the latter, and that, depending on the geological and climatic situation, it is derived additionally or predominantly from the sheet-wash in the catchment area.

The composition of the suspended load is as complex as its origin. Whereas the particles of the bedload represent a compact combination of minerals, the source material of the suspended load has been largely disintegrated into grains of the individual minerals. Depending on the original grain size of its minerals, the source rock will disintegrate rather rapidly into its components below a certain size. Very fine-grained material survives in distinct rock fragments down to the smallest particle sizes. In general, however, at grain sizes of $1-2$ mm a rock may be said to have been completely broken up.

In addition to the determination of the grain sizes, the investigation of the suspended load also entails a microscopic analysis of the mineralogic-petrographic composition of the particles. Of interest is not only the composition of the grain mixture, but also the degree of rounding of the individual particles. The harder a mineral grain (e.g., quartz) is, the more angular it will remain, and if this is a main component, it will contribute considerably to the wear exerted on the rotor blades of turbines in hydropower stations. The mineral grains are counted according to the usual statistical methods whereas for the degree of rounding a number of approximative methods, found amply sufficient for this purpose, may be employed (Sect. 4.2.2).

As has been already said, the ease of transportation by the flowing medium and the ensuing thorough mixing render an exact determination of the derivation and the respective classification of the various minerals rather difficult, particularly so, as index minerals derived from a particular area are usually rare. In sedimentary petrography heavy mineral analyses are carried out which permit the identification of certain source areas. This method is helpful for the paleogeographic interpretation of ancient river and marine sediments. But it is not always applicable to the suspended material collected directly out of the flowing water. In contrast to this it is quite common for deposited sediments transported previously close to the river bottom, and is used here, e.g., for geochemical prospecting (Sect. 4.2.3). Mineral components of the suspended load are quartz in a multitude of shapes, feldspars, mica flakes, calcite and dolomite particles, small rock fragments, and organic compounds. There is also material of anthropogenic derivation from refuse, sewage, rubble, etc. Included are furthermore industrial remnants from ore treatment sludges, slimes from clay and sand pits, as well as clay minerals. The bulk, however, is made up of quartz, feldspar, micaceous minerals, or carbonate, depending on the source area considered.

The mineralogical composition of the suspended load is only rarely illustrated graphically. The common bar diagrams for the bed load are used, as well as the so-called rock or mineral spectra in which certain minerals or related groups of minerals are plotted. The ordinate gives the cumulative percentages and the abscissa the grain size scale. The quantitative relation of the individual mineral or the group of minerals is then shown in the form of a curve.

According to Lichtenhahn (1977a), knowledge of the suspended load transport and its composition is of importance for the following aspects:

- silting-up of abandoned meanders,
- formation of deltas in lakes and the sea,
- silting of bays,
- silting of reservoirs,
- flushing of deposited suspended load in reservoirs (endangering fish downstream),
- deposition of clays on flooded areas (either destruction of crops or fertilization),
- lateral sealing of a river bed against groundwater,
- water intake for public water supply,
- wear of turbines (sand retention devices).

4.3.4 Distribution of Suspended Matter

Water containing suspended matter has a higher density than clear water. If C_s denotes the concentration of suspended load, i.e., the amount of suspended load per unit of space in the liquid, then the density of the mixture will be

$$\varrho = \varrho_w - \frac{\varrho_w C_s}{\varrho_s} + C_s \ . \tag{82}$$

Across the section of a river the density is not constant, as the suspended particles tend to deposit under the influence of gravity and thus lead to higher concentrations near the bottom.

There are a number of theories dealing with the vertical changes in concentration. The best-known one is the diffusion theory by Rouse (1936) which is based on Prandtl's concept for turbulent exchange. It will be briefly discussed here.

The movement of water in rivers will always be turbulent, i.e., a random and irregular mixing movement is superimposed onto the main flow, leading to an apparent increase of the viscosity. Based on a logarithmic distribution of the velocity (Chap. 3.1.2), according to Dillo (1960) and Burz (1967), the following considerations will be valid: All rising water particles move into zones of higher velocity, all sinking particles into zones of lower flow velocity. As a consequence, the component of the velocity parallel to the main flow in a certain water layer will be increased when particles enter this layer from above, and it will be reduced when particles rise into it from below. These accelerations and retardations lead to a transfer of forces which was derived by Prandtl from the law of impulses. According to this, the apparent shear stress $\tau' = A(\delta v/\delta z)$ is a function of the so-called "exchange" A, a locally undefined value with the dimension of a viscosity, and of the vertical distribution of the average velocity in the direction of the main flow.

The vertical distribution of the suspended load can be presented in a manner similar to the exchange of impulses. The turbulent exchange brings suspended material from zones of higher concentration near the bottom into zones of lower concentration. This upward transport is counteracted by the downward velocity component of the turbulence and the settling velocity of the suspended load, leading to the situation that the concentration close to the bottom will always be greater than that of the higher zones. Provided that the distribution is stationary, an equilibrium is established thereby and maintained between the settling of the particles under the influence of gravity and the turbulent cross-movement of the suspension-laden water.

From a rather involved mathematical consideration based on a triangular distribution of the shear stresses, Rouse arrived at a ratio of concentrations of the suspended load at depth z and a:

$$\frac{C_{Sz}}{C_{Sa}} = \left(\frac{h-z}{z} \cdot \frac{a}{h-a} \right)^{\alpha} , \qquad\qquad (83)$$

with $\qquad \alpha = \dfrac{\omega}{0.4 u*}$

and $\qquad u* = \sqrt{ghJ}$.

The settling velocity w can be taken from Fig. 4.1 or calculated from Stokes' law [Eq. (30)]. Equation (83) is not fully satisfactory as it leads to infinitely large concentrations immediately near the bottom ($z \to 0$). Therefore, close to the bottom a layer with the thickness $a = 0.05 h$, the so-called bottom layer is commonly excluded from the considerations. In Fig. 4.46 the distribution of the concentrations is shown for different exponents α.

From a similar approach, Müller (1973) developed the modified equation

$$\frac{C_{Sz}}{C_{Sa}} = \left\{ \frac{a}{z} \cdot \left[\frac{\sqrt{h}+\sqrt{h-z}}{\sqrt{h}+\sqrt{h-a}} \right]^2 \right\}^{\alpha} , \qquad\qquad (84)$$

and could show that the distribution curves calculated with it exhibit a better agreement with the extensive experimental results published by Vanoni (1941) than with the curves calculated by Rouse.

In addition to the diffusion theory, the gravity theory of Welikanov (in Bogárdi 1974) has become known, which considers the transport of the liquid and the solid phase of the suspension separately. From a complicated derivation Welikanov found the equation

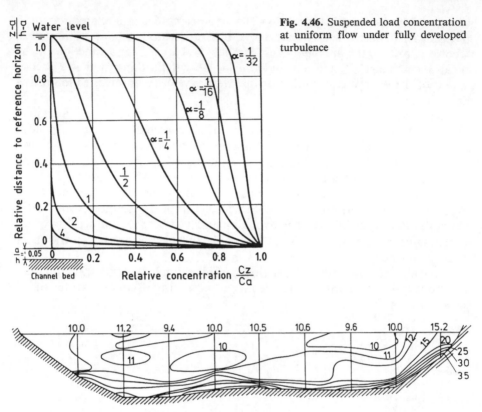

Fig. 4.46. Suspended load concentration at uniform flow under fully developed turbulence

Fig. 4.47. Distribution of suspended load in $g\,m^{-3}$ in Danube at Engelhardtszell on 15. 4. 1957. (From data of Donaukraftwerke Jochenstein AG)

$$\frac{C_{Sz}}{C_{Sa}} = (1-\eta)^{m\beta} \ , \tag{85}$$

in which η is the relative depth, m a function of the relative bed roughness, and β a factor containing several parameters.

Several recent theories communicated by Bogárdi attempt to clarify with considerable mathematical effort the rather involved problem of the distribution of the suspended load. The practical importance of these methods is rather limited as the actual distribution of the concentrations was found by measurements to be subject to considerable irregularities and fluctuations. This applies not only to the vertical but also to the horizontal distribution of suspended load. For the latter there are so far no theoretical solutions.

The observation that the determined concentrations virtually always differ from the theoretically expected ones can be explained by two main reasons: Firstly, the grain sizes of natural suspended load range from sand to the finest silt and clay particles and thus only imperfectly fulfill the conditions of Eq. (83), which assumes a uniform grain size. Secondly, the transport of suspended load will only rarely be homogeneous, the suspended particles frequently forming cloud-like condensations within which the concentration will be larger than in their vicinity. The sample collected at a certain point at a certain time can thus supply a value completely different from that of a sample taken only a short period previously at the same place.

From the vertical lines of the overall cross-section measured, an instructive picture of the distribution of the suspended load can be obtained, similar to the presentation of the isotaches in the case of stream gaging. As an example the result of the full-range measurements of the suspended load of the Danube at Engelhardtszell on May 14th, 1957 is presented in Fig. 4.47.

5 Channel Geometry

5.1 Network Evolution as an Interconnected System

Four large complexes form the natural bases of the formative processes of river beds:

- tectonics as the entirety of the horizontal and vertical movements of the crust controlled by endogene forces;
- the lithology as a collective description of the mineralogical composition of the surface of the earth including the soils;
- the climate as the sum of all atmospheric phenomena;
- the vegetation.

With these complexes nature maintains the equilibrium of many interdependent systems, rivers being considered as one of them. From these natural bases two important transportation processes are derived: run-off of water and the transportation of sediments. These phenomena in turn are the main controlling factors for the dimensions of river beds, i.e., for the channel geometry (Fig. 5.1, cf. also Morisawa 1968, 1985, and Allen in Gregory 1977).

Every river is seen as a three-dimensional body. In order to gain an insight into the laws of river bed evolution, it is necessary to accept the elements of channel geometry, i.e., plan view as well as longitudinal profile and cross-section, as parts of the whole, and to understand their mutual dependence. The elements can be taken as the members of a feed-back or closed circuit which influence each other in shape and magnitude. They are also part of an "interconnected system" which is furthermore controlled by the external factors, discharge and sediment load. The directions of these influences are shown by arrows in Fig. 5.1. Save for the vegetation, the human influences − disregarding hydraulic engineering − extend only to the processes of transportation, which in turn influence the shape of the beds. Vegetation, as one of the natural conditions, takes an intermediate position but on the other hand is strongly modified in its extent by human activities. This can lead to considerable increases in discharge and sediment yield, and even influence on climates cannot be excluded where large-scale deforestation is practised. On the other hand, long-term climatic changes can accelerate forest-clearing programs.

For methodical purposes each aspect of channel geometry will be considered separately. However, the limitations set by this subdivision have to be neglected at times in order to illustrate the mutual interactions of the bed-forming forces in the development of certain bed forms of rivers.

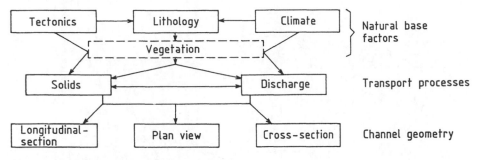

Fig. 5.1. Controls on the morphology of river beds

5.2 River Channel Patterns

5.2.1 Descriptive Parameters of Channel Development

To characterize the longitudinal development of rivers, a number of dimensional factors can generally be given. The official river distances usually take the river regulations into account and start from the mouth upward, as this is the only point that can usually be defined with sufficient accuracy. This method, however, has not been uniformly applied. There are exceptions, for example the Rhine, which is measured from Constance in a downstream direction (Kuhr 1972) or the Danube with its mutually opposing methods. In the latter case there are three different distance sections: In Württemberg down to Ulm, thence in Bavaria downstream to Kelheim, and then the uniform river kilometer notation from the mouth near Sulina up to Kelheim. Old and new kilometer points do not coincide, an agreement on the international river kilometer system for the complete course having been reached here only very recently. The course of a river derives from the adjustment of the running water to the geological substrate and the morphology of the area concerned, as well as from the active sculpturing of the area by denudation and accumulation. It appears reasonable to express the relationship between the source area and the river course as well as its total length by a number of ratios. For this purpose the direct line c from the source to the mouth is deduced from the length of the river l_F and divided by c (Fig. 5.2). It thus is

$$e_F = \frac{l_F - c}{c} \; .$$

The expression e_F is called river evolution (cf. Wundt 1953; Keller 1962, etc.). This relationship is valid for the entire length of the river, but can also be applied to its individual sections. In this case c stands for the distance between the points A and B. Figure 5.2 furthermore takes no tributaries into account which themselves can influence the topographical development of the main stream. Consequently in many cases, especially with larger rivers, it is more convenient to consider individual subsections which show typologically similar developments.

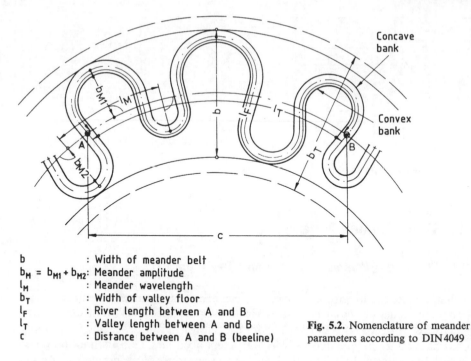

b	: Width of meander belt
$b_M = b_{M1} + b_{M2}$	: Meander amplitude
l_M	: Meander wavelength
b_T	: Width of valley floor
l_F	: River length between A and B
l_T	: Valley length between A and B
c	: Distance between A and B (beeline)

Fig. 5.2. Nomenclature of meander parameters according to DIN 4049

The river evolution is large for rivers which flow in a large circle such as the Orinoco, Niger, and Congo (Zaïre) and it is moreover independent of the size of the catchment area. It is smaller when the river flows in almost a straight line such as the Po. For the Rhine from source to mouth Wundt (1953) gave a value of about 0.9. For the short section from Bonn to Cologne, as an example for parts of the course, the value is only 0.38. The latter value is already influenced by the regulations of the course of the Rhine.

The separate evaluation of river sections is particularly useful for meandering rivers, but fails for highly braided ones. Instead of the straight line c the curved line l_T (valley length) is taken for meandering rivers (Fig. 5.2) and the expression

$$e_L = \frac{l_F - l_T}{l_T}$$

is called course evolution. The valley evolution is defined by

$$e_T = \frac{l_T - c}{c} \; .$$

A few other factors such as meander length, l_M, and meander width, b_M, result accordingly. The meander bed or the width of the meander belt are not always easy to define. Frequently an elevated bank is developed, such as the rather characteristic lower terraces of the Würm ice age (e.g., Wisconsin, USA).

In this case, l_T is the median line between these elevated banks. The use of the sides of the valley is problematic with incised meanders (Wundt) or complicated valley meanders, as here even more points of limitation are present (Sect. 5.2.4.1).

Wundt (1953) has pointed out that in older reports (e.g., Gravelius 1914) for river, course, and valley evolution the value for c and l_T are not deducted prior to division by c or l_T and that the respective expressions read as $l_F : c$, $l_F : l_T$, and $l_T : c$ instead. The resulting values then are all higher by 1.0.

For the determination of river and course evolution the map scale is of importance. It also depends on the scope of the study, on what scale the investigation has to be carried out, i.e., whether individual bends have to be considered, as would be the correct approach, or whether for surveys of larger rivers the length of the valley will suffice. The lack of accuracy of maps used in third-world countries frequently demands some give and take. For detailed investigations it is of importance whether the channel line of the river, the connecting line between the points of highest surface velocity, a middle line to be agreed upon, or the thalweg are used. The start and end of a meandering stretch is also not always easy to define. All documents used, as well as the method applied, always have to be quoted.

It has become customary to subdivide rivers into straight, braided, and sinuous reaches (Leopold and Wolman 1957), although the visual approach could cover also other configurations. Several authors consequently use sinusoidal undulation and anastomosis as additional means of subdivision. Schumm (1960) used the term anastomosing for a braided channel, the islands of which have stabilized their position or were stable from the start. They are no longer subject to continuous shifting and thus form, as it were, a stabilized braided system. This is not without contradiction in itself, and Weber (1967) voiced his opposition thereto, although in a different context.

Miall (1977) took over this expression, and Morisawa (1985) illustrated the various outlines tabulated by Miall, i.e., straight, sinuous, meandering, braided, and anastomosing. However, as all shapes can be derived from three basic ones, in the following the subdivision into straight, meandering, and braided will be used. In certain cases the straight shape can be considered as a special form caused mostly by tectonic forces.

5.2.2 Straight River Channels

In nature there are no geometrically straight rivers. Channels are thus called straight if they show a very small river evolution over a certain distance, and in general give the impression of pronounced elongation. Leopold et al. (1964) stressed that rivers rarely flow in a straight channel for more than about ten channel widths and that the term straight also covers all somewhat irregular, slightly winding, nonmeandering rivers, this being basically identical in German usage to a very small river evolution.

Two main factors govern the straight nature of a channel: a steep gradient and a narrow course caused by geological and morphological influences.

Elongation due to steep gradients is widespread in geologically young mountain ranges with a corresponding energy of relief and in general in tectonically active regions. Each scour basically has this character, whether carved into crystalline rocks or limestones, on the slope of an ash volcano or on a mine dump. Almost all torrents and many mountain creeks belong to this group (Chap. 6.3.2.1), short reaches being typical for them.

Straight reaches furthermore frequently originate from geological and morphological factors during epigenesis and antecedent formation of valleys (Chap. 7.1.2.2) as long as the gradient is steep enough. Regressive erosion results in the most striking examples. It expresses itself in the form of gorges and ravines which do not have to be exactly straight, as well as in the form of valley reaches of different widths, where harder rocks form the substrate to be incised by the river. Bends here are only slightly rounded and abrupt changes of direction predominate.

Straight river reaches are rare, even when the river bed follows its own alluvials. They appear to be stable only if the bedload transport is small and the sediments in the channel are only very little moved by the flow. The thalweg does not run regularly along the center, but oscillates between the banks. In the longitudinal direction bars alternate with scours, giving the river the appearance of a chain of elongated pools in contact with each other in the transitional parts, the so-called crossings or riffles. The low-water level has its biggest gradient on the riffles and its smallest over the pools.

Any undulations are smoothed out with increasing water level, until eventually − at the highest levels even over longer stretches of the river − an almost plane run of the longitudinal section of the water level develops. During floods more bedload is transported over the pools than can be accommodated by the subsequent riffle. Consequently the latter is raised in elevation whereas the pool is deepened. During smaller discharge, however, this process is reversed as channels are eroded into the riffles and the resulting material tends to fill the pools.

In small-scale maps minor morphological features of the river course are frequently suppressed to such an extent that the impression of a straight course results, and the river as a whole is then classified as of the straight type. One example is the Nile, especially north of the geologically controlled large bend of Abu Hammed and Debba.

5.2.3 Braided River Channels

Rivers of this type are widely developed and are specifically encountered in areas with strong bedload transport. Medium to large gradients are a prerequisite. Initially the following situation is encountered:

In contrast to the straight channels, there are no confined beds with unchanging banks. The river is split into numerous diverging channels which reunite again and change their appearance with every major flood. While one channel is being filled up, the water scours a new path right next to it, until the next flood leads to its starvation and widens another, previously unimportant channel to the main arm with the then strongest discharge. Gravel bars and islands devoid of vegetation are spread out between the channels and are continuously being relocated.

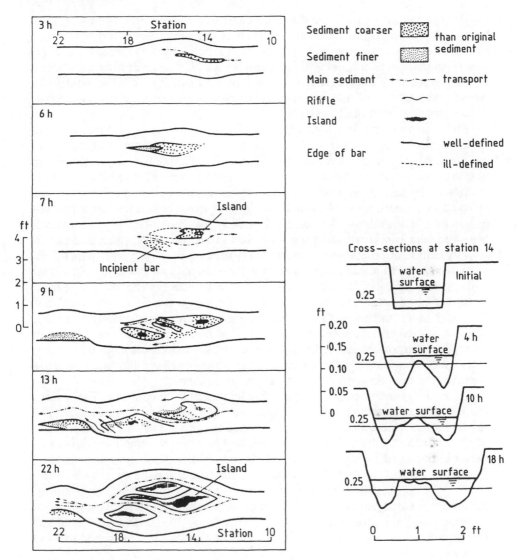

Fig. 5.3. Development of braids in a flume. (After Leopold and Wolman 1957, 1964)

Any incipient vegetation is carried away by the next flood. The wide band almost devoid of vegetation which is required by the river for the continuous modification of its intertwined arms from above looks like an irregular braid work. The greater part of the wide gravelly bed disappears only during floods under the masses of water discharged.

What are the causes leading to the braiding of a channel? The above-mentioned parameters of gradient and bedload transport appear to be the controlling factors. Present field data and experiments permit a more differentiated approach. With the aid of laboratory flumes attempts have been made to gain an insight into the start of the braiding of a channel. Leopold and Wolman (1957) have published interesting observations. Their experiments were based on the simple fact that a river starts to split up by flowing around one or more islands. How then do these islands originate? This process is the initial stage of each separation, whether induced by discharge, bedload yield or gradient. Figure 5.3 illustrates the evolutional sequence of such an experiment which is in agreement with numerous observations on natural channels.

For these experiments a layer of an only moderately sorted mixture of coarse sand was placed into a flume and a straight groove was drawn into this moist loose sediment. About 3 h after the gradual addition of water, the first slight deposition of grains somewhat larger than the average was observed. This resulted in an inhibition of the discharge, leading to an increased accumulation initially of coarse material. In general, for continued deposition one needs only a few larger pebbles or boulders. The authors also observed the well-known ball-bearing effect which helps finer material carry along coarser or very coarse particles which at the given velocity could not otherwise have been transported on their own (cf. Chap. 4.2.6).

After 7 h the sand bar formed was so well developed and steep that individual grains started to roll off its side, causing small turbulences in the water, which led to further obstacles. The upward growth of the bar to the water level displaced the water more and more to the sides into newly forming deviating channels, where it started to encroach on the outside bank and to erode the bottom to satisfy its transport capacity. In this way the cross-section was enlarged, almost every time accompanied by a lowering of the water level, and leading, at least in parts, to the establishment of the sand bar as an island.

In nature these processes are undoubtedly much more complex. The various shapes are particularly clear on so-called "sandr" plains such as in Iceland. The sandrs present spectacular examples of braided channels. They are a combination of alluvial cones and braided river systems, and in recent decades have repeatedly been the object of intensive studies (Hjulström et al. 1954; Thorarinsson 1960; Krigström 1962; Church 1972). A number of observations, especially by Krigström, will be presented below as they appear to be of general applicability. Krigström pointed out that parts of his observations are in astonishingly good agreement with the experimental results of Leopold and Wolman (1957). Braiding of channels on sandrs takes place in two ways which can be considered to be at least as related:

1. The sandr rivers are at times overloaded with bedload. Consequently they aggrade their bottoms to such an extent that the channel literally plugs itself and the water is displaced laterally. This case, which originally was thought to be the cause for braiding in all rivers, is almost entirely restricted to flood events. These do occur regularly on sandrs, but are not particularly frequent.
2. The second, more frequent type of braiding is caused by sand or gravel bars, which have also been described from flume experiments by Leopold and Wolman. It is the type occurring at average to low discharge rates, or generally at receding water levels. Its natural precondition is a certain amount of bedload transport, even if only of sand. As the gravel or sand bars on a small and large scale show identical shapes, Krigström intentionally omitted a scale from his

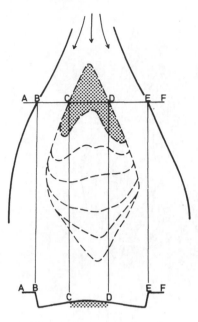

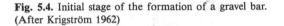

Fig. 5.4. Initial stage of the formation of a gravel bar. (After Krigström 1962)

illustrations. A slight widening of the channel resulted in a reduction of the flow velocity of the water, this widening being associated with a change in gradient. It does not, however, appear to be necessary – any reason for the deposition of bedload carried along by the water will suffice, as reported by Leopold and Wolman. The aggradation of the bed is frequently more pronounced in the center of the channel, a situation which appears to be another reason for the sedimentation of the bedload.

In Figs. 5.4 and 5.5 the evolution of gravel bars is presented from field observations. The initial stages of bar formation were difficult to follow due to the turbid nature of the melt water. Figure 5.4 illustrates the probable evolution from an embryonic stage. At the beginning, the bar advanced more rapidly at the outside points, leading to an arrowhead shape as an intermediate stage. This was gradually transformed into a spindle-shaped streamlined body called spool bar. As at the same time this body grows upward, its water cover becomes thinner and eventually the water level is broached. This can be caused by any recession of the water level.

Erosion is now increased in the side channels and the water level drops more rapidly so that the lower point of the gravel bar and its lateral extension emerge from the water initially on the upstream side with an inverted arrowhead shape (Fig. 5.5a). The remaining flow of water across the bar erodes it into various, again partly spindle-shaped small bodies as the base level for the erosion at the downstream end is at a lower elevation. Despite this, the parts emerging first need not be the highest ones, but usually the upper end will be highest. At a further drop of the water level and/or further deposition of bedload, the gravel bar is rarely overtopped again by water (Fig. 5.5b): an island has formed (Fig. 5.5c).

This shape of a short wide spindle is observed worldwide, and is so characteristic that it can be used as a symbol for the braiding of rivers.

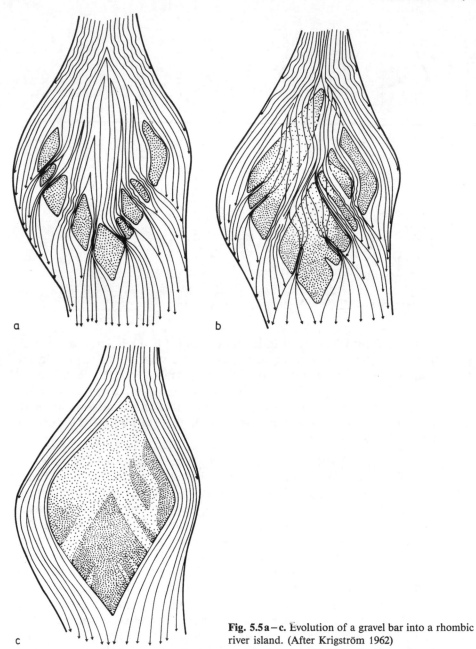

Fig. 5.5a–c. Evolution of a gravel bar into a rhombic river island. (After Krigström 1962)

Similar shapes were observed at the confluence of two channels. In Fig. 5.6 only one of the two channels was considered as bearing a notable amount of bedload. This situation leads to the formation of gravel bars similar to those shown in Fig. 5.5. In case one of the tributaries plugs itself or if one of them (a in Fig. 5.6) erodes more strongly, that channel in branch b will take over the

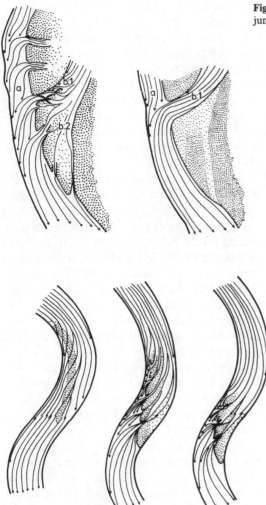

Fig. 5.6. Development of a point bar at the junction with a tributary

Fig. 5.7. Development of a gravel bar below a river bend. (After Krigström 1962)

total discharge which achieves the connection with branch a most rapidly by erosion. This will usually be the one requiring the shortest distance – a bar is formed. This can be overflown again at rising water level and lead to the deflection of the line of highest velocity of the channel from its outer edge.

Frequently these bars, as the point bars formed on the convex bank of a river bend, are shifted into the course of the river. They form bars which characteristically extend obliquely across parts, or at pronounced bedload discharge across the entire width of the river, frequently connecting with the next bar. They can considerably support the river bed in the upstream direction (Fig. 5.7).

Experiments as well as field observations have confirmed that there is no need for large bedload transport or water discharge to initiate the braiding of a river. The braided river represents a stage of equilibrium which may be maintained over an astonishingly long period, provided that discharge and supply of bedload do not change markedly. It is thus a type separate from the meandering rivers and not only a subtype of the latter. Braiding is, as it were, the response of a river to the energy of the relief and the amount of bedload to be transported.

Compared to a nonbraided confined channel, an equally large but braided one shows a steeper gradient. This observation has led to the concept that a confined channel is more characteristic of lower reaches in lowlands and that during braiding the river attains the steeper nature of the upper reaches in order to compensate for the increase in gradient or to transport the bedload through the steeper reach (Leopold et al. 1964). This explanation, however, is in conflict with a number of morphological and hydraulic criteria.

A special type of branching is the redeposition or rebedding reach. When there is a balanced relationship between average bedload discharge and transport capacity (Chap. 4.2.7), neither deposition nor erosion will take place in the long term. The redeposition reach is characterized by the fact that the bottom elevation can vary within a certain range, depending on the supply and removal of bedload, but on average is rather stable. Redeposition reaches frequently form upstream of narrows, in mountainous areas sometimes even several in succession. The narrows are of natural origin, such as rock ledges, but there are also artificial obstacles such as groynes, bridge pillars, bottom ridges, weirs, etc. The space behind these can be considered as a retention area for bedload, out of which the river will transport just as much through the narrows as its transport capacity at the time will permit. During floods the volumes can be huge.

In earlier years numerous Alpine or pre-Alpine rivers such as the Salzach, Isar, Lech, and Iller presented examples of this situation. The original morphology is barely recognizable any more because of the modifications carried out during the regulations of the last century. Only in older maps is the original state still shown, frequently together with the new shortened course. This illustrates how much man has modified the river landscapes.

The redeposition reach above narrows can be considered as a strongly stretched, laterally confined alluvial cone. Examples of this type are virtually all inner-Alpine rivers with stronger gradients such as, for example, the Inn. These are not only constricted and displaced by numerous small lateral debris cones, but are themselves flowing, so to speak, over a very much elongated alluvial cone. Regulation of the rivers has greatly obliterated the original picture.

Behind large lateral cones the river had frequently become dammed up and formed a chain of redeposition reaches. This results in stepped valleys.

Even on large, and particularly little inclined alluvial cones such redeposition reaches can be formed. On very active cones the entire surface is affected, i.e., depending on discharge and supply of bedload material the creek or stream oscillates between the banks. The fan in its entirety then represents a braided section. Sandr plains of large glaciers are actually nothing else than this.

Disturbances of a braided reach can be caused by a variation of one of the two basic parameters. Changes in the supply of bedload are most impressive in this respect. If this supply is reduced for any reason, braided rivers quickly lose their characteristic features. Degradation − less frequently bank erosion − starts to prevail and the network of braided sections is replaced by a stretched chan-

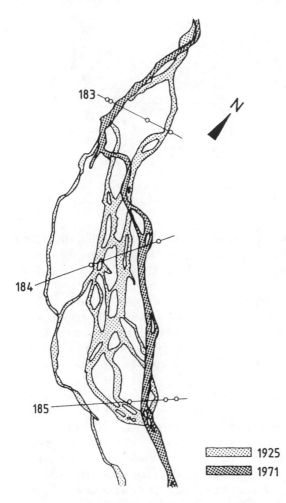

Fig. 5.8. Reversal of the Isar from braided to stretched course in the Ascholding flood plain

183

184

185

▒▒▒▒ 1925
▒▒▒▒ 1971

nel that cuts itself into the bottom and causes the former side channels to fall dry.

An example of such a reversal of a braided river to a stretched course is given by the Isar in the Ascholding and Puppling flood plains, a region that until a few decades ago was one of the last riverine wilderness landscapes not yet affected by man (Scheurmann 1973). The several hundred meters-wide bed was cut by numerous individual arms and channels. On the laterally adjoining terraces, which had been formed as a result of the gradual post-glacial degradation of the river bed and are very rarely flooded, a highly varied flood plain vegetation has established itself. A fundamental change in the morphology was introduced by the large structures built in the upper reaches of the Isar such as the power plants of Walchensee and Bad Tölz and the Sylvenstein reservoir with the ensuing changes in river regime. The reduced supply of bedload and the absence of heavy floods inhibit the transposition of the bedload required for the maintenance of the braided river system. In order to satisfy its transport capacity, the Isar started

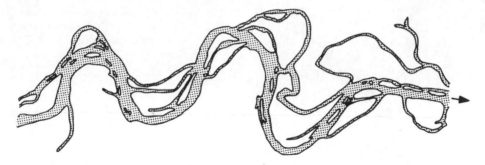

Fig. 5.9. Channel system of the lower Isar above Plattling prior to regulation

degrading its own bed and formed a stretched main channel, while the lateral branches started to dry up and deteriorated rapidly. At the same time the willow-tamarind flora started to encroach onto the gravel bars, leading to their solidification (Fig. 5.8).

The reversal of braided rivers by regulation or retention of bedload is a widely developed phenomenon. It affects many rivers of the civilized world and does not spare other sections either. There is virtually no other landscape so much affected by man's activities as are rivers and their valleys. On the other hand in past centuries, they were always the most important areas for settlement and the main arteries for transportation.

When the dynamics of a river diminish with decreasing gradient, braiding will gradually disappear, giving way to the meandering channel type. The change is rarely abrupt, but usually passes through intermediate stages in which the original tendency still prevails to some extent. In the braided network, a meandering main channel starts to govern the appearance of the river, while the other branches deteriorate into unimportant side lines. The Isar again offers an instructive example. Prior to the regulations of the 19th century almost its entire course in Bavaria could be considered as a chain of braided reaches, whereas in the less inclined lower stretch, especially between Dingolfing and the mouth, characteristic intermediate stages were developed, marking a transition to the meandering type (Fig. 5.9).

Wundt (1949) compared the change in bed morphology with the establishment of a state of quiescence between erosion and accumulation and gave the Rhine below Basel as an example. Already in the reach down to Breisach at a gradient of about 1‰, the untamed river prior to the Tulla regulations showed a tendency to small meanders, but "self-plugging" and frequent relocations of the course did not allow the development of real loops. Only below Strassburg, where the gradient is much smaller, could an equilibrium establish itself, leading to well-developed loops (Fig. 5.10).

The transition between the braided and the meandering type is difficult to fix exactly, as the physical laws of channel geometry are not sufficiently understood. Because of this limitation, empirical rules are applied which have been derived from field observations and laboratory experiments.

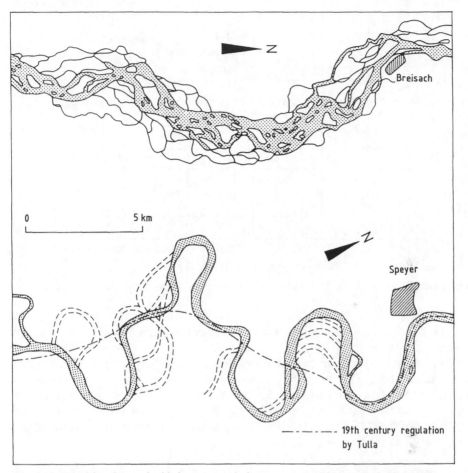

Fig. 5.10. Transition from a braided to a meandering course: the Rhine at Breisach and Speyer

According to Leopold and Wolman (1957) the boundary is controlled by the gradient of the bed in relation to the bank-full discharge. In metric units the relationship is

$$J_s = 0.012 Q^{-0.44} . \tag{86}$$

A rapid estimation as to which channel type should be developed is possible with the aid of Fig. 5.11.

If the actual gradient is larger than the one obtained with Eq. (86), braiding has to be expected, otherwise meandering should be expected. The coefficient (0.012) and the exponent (−0.44) include all parameters controlling the channel formation save for the discharge. Consequently, they cannot be universally applicable constants, as pointed out by Zeller (1967). It is, however, known from the regime theory (Sect. 5.6) that for a certain drainage basin the two values will vary only slightly. The approximative nature of the approach used by Leopold and Wolman can therefore be tolerated. Within the range described by Eq. (86) there

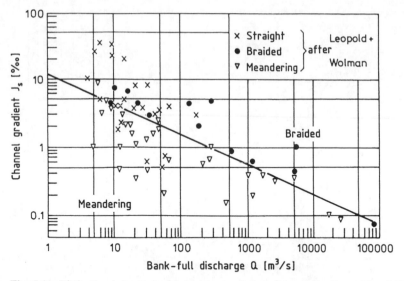

Fig. 5.11. Distinction between braided and meandering rivers. (After Leopold and Wolman)

will be flowing transitions, as shown by the example of the lower Inn mentioned above.

If river systems with fundamentally different hydraulic and bedload properties are compared, Eq. (86) might fail, as the fixed values in this case would have to be replaced by functions for the controlling parameters. The indeterminate nature of the boundary is also suggested by the large spread of the points marking straight channels in Fig. 5.11. No theory allowing a better definition of the boundary has yet been found.

5.2.4 Meandering River Channels

Sinuous river reaches are usually referred to as meanders. The name was derived from the river Menderes in Anatolia, which as Maiandros had already attracted the attention of the ancient Greeks because of its remarkable loops. The term meander has since become a term of classification for a certain type of pattern, without its features becoming more defined than by just a rough visual designation.

It was subsequently found that there are various types of meanders which indicate different underlying causes. Usually they are subdivided into free or river meanders which develop in their own alluvials, and incised meanders or meandering valleys, the sinuous nature of which depends on the topography of the countryside. There are furthermore several intermediate or mixed types. Amongst these are the meandering rivers called undercalibrated by Zeller (1967). These are flowing in a relic valley actually too wide for them, such as, for example, the Altmühl following a paleovalley of the Danube between Dollnstein and Kelheim. Other generic terms such as active or inactive meanders indicate the tendency of the river in the first case to relocate its bed continuously, whereas in the second case its configuration remains unaltered over a considerable span of time.

5.2.4.1 Meandering Valleys, Incised Meanders

The large sinuous valleys deeply incised into the countryside form a rather conspicuous phenomenon of river morphology. Beyond doubt they have developed from rivers, but because of their quite different configurations, a number of genetic theories have been proposed which will be described below. They furthermore represent an interesting aspect of the history of science in this subproblem of morphological and geological research. It is frequently not possible to separate the theories from each other clearly, and thus the terms applied tend to express only the basic concepts used. A compilation of the various theories is given by Mahard (1942), Baulig (1948), Ermeling (1951), and Troll (1954).

The rather diverse opinions on the formation of meandering valleys in the literature have led to a host of differently applied terms, and at times a certain term was used for two different situations. As they are frequently merely of historic interest, only the presently accepted terms will be mentioned. Generally recognized and used are the terms "incised meander" or "meander valley".

In English-speaking countries the following terms are applied: Intrenched meander, incised meander (predominant) as well as valley meander or meander valley, the latter term being preferred in literature. Other terms are entrenched meander and ingrown meander, describing the difference between meander valleys with symmetric slopes (entrenched) and asymmetric slopes (ingrown). Inclosed meander is a rarely used term. In French-speaking countries "méandre encaissé" is used, "méandre imprimé" and "méandre sculpté" being outdated.

Inheritance Theory. Ramsay (1874); de la Noë and de Margerie (1888, in Ermeling 1951); Davis (1893, 1903, 1909, 1923); Philippson (1924); Baulig (1948); criticism by Scheu (1909); Schad (1912); Rich (1914); Wagner (1919, 1929); Engelmann (1922).

According to Ermeling (1951) the French authors de la Noë and de Margerie (1888) in their article *Les formes du terrain* for the first time expressed the concept that incised meanders (méandres encaissés) formed from freely mobile river meanders (méandres divagants) during uplift of the countryside. However, Ramsay (1874) had already voiced these ideas much earlier in a description of the Rhine. The main protagonist of this theory was W.M. Davis, who used it in the development of his cycle theory. His illustrations of the morphological cycle were adapted in many textbooks and at times represented the "ruling" opinion. The inheritance theory equates meander valleys and river meanders, the former being simply river loops fixed during uplift so that they could no longer develop without impediment. Philippson (1924) consequently called them forced meanders, similarly to Rich (1914) and Inglis (1949). He distinguished between the real forced meanders with the slopes on either side being equally steep (symmetrical) which are directly inherited, and the free meander with typical concave and convex banks. This has developed further during the process of incision, and inheritance was only the starting point.

Twin-Cycle Theory. Davis (1893, 1903, 1909); Morisawa (1968, 1985); criticism by Leopold et al. (1964).

This represents a modified version of the inheritance theory and postulates that during renewed incision for whatever reason, a river will retain its channel geometry attained by then, this representing the first cycle. This applies to all channel shapes from simple sinusoidal undulations to mature river meanders. The incised valley, which in all main parameters conforms to the geometry of the original channel, then represents the second stage or second cycle.

The valley meander displaying a symmetrical cross-section corresponds to the forced meanders of Philippson, being considered characteristic for this process.

If the incised meander is asymmetrical, i.e., a slope meander, the shape of the channel will be developed further during the process of incision. It is then an original first-cycle meander which is still to start on the second cycle.

The morphologies described so far are restricted to only slightly inclined tablelands. There are still major uncertainties in finding a plausible explanation for the derivation of new shapes from the earlier ones; and it has, moreover, not yet been shown unequivocally how the symmetrical slopes have been formed. As a consequence, Leopold et al. (1964) do not consider a special twin-cycle theory necessary, even if in certain cases such minor special shapes might be present. They give for an example a glacial creek that has cut through a layer of ice into the underlying solid bedrock (Leopold et al. 1964 p. 314) and thus do not a priori exclude the possibility of inheritance.

Climatic Morphology and Valley Meanders. Wagner (1919, 1929, 1955, 1960); Hol (1938); Dury (1953); Kremer (1954); Troll (1954); Leopold et al. (1964); Schaefer (1966).

The at times conspicuous discrepancy between the size of the large sinuous valley and the river now flowing in it being much too small to have produced this shape requires an explanation. The presence of much larger discharges during earlier periods has to be assumed by way of an explanation for the considerably larger erosive processes. This does not, however, apply to all rivers. In some meander valleys the river even today occupies the complete width, actively continuing with the process of degradation. This climatically controlled type of meander valley was first described from deeply eroded shield areas. It is, however, not restricted to these and may also be found in regions of Quarternary glaciation.

The difference can be ascribed mainly to the geology of the substrate. In regions of northern hemisphere glaciation of Europe and North America as well as in the Alps, the meander valleys were initially formed in soft moraine material and gravel plains. The resulting morphology was then destroyed by the next glaciation, unless it had been cut into harder and thus more resistant bedrock. In these cases and in ice-free regions they could survive longer and it was here that they were first described.

These valleys may be taken as belonging to the slope meanders in old shields and table lands as well as in glacial gravel plains. They thus have not only cut themselves in, but have suffered further developments. For detailed examples reference is made to Wagner (1955, 1960).

Regarding the rivers of southern England Dury (1953, 1964) found that the length of a valley meander usually is equivalent to ten times the length of the meander of the river presently occupying the respective valley. He arrived at the conclusion that meander valleys were generally formed only in periods of higher discharge rates, such as for example during the Quarternary.

Hol (1938) and Troll (1954) observed that at the fossil melt water gates of the Alpine foreland glaciers, meanders cut down to 50 m into gravel layers, moraine material, and partly into soft Tertiary rocks, thus forming innumerable terraces, oxbow lakes, cut-offs, etc. which can be equated with meander valleys. Hol therefore called them meander terraces, equivalents of the incised meanders. They originated during very short periods of high discharge of melt water from the glaciers and at rather high gradients on the apex of the debris cone close to the face of the original glacier. According to Troll, they could represent the initial stages of meander valleys in solid bedrock. The examples described, especially those from the Inn and Lech in Bavaria, presently show almost no further development and due to their special conditions of formation fail to yield generally applicable solutions. However, the theory of climate-controlled formation of meander valleys remains correct in as far as it covers the possibility of widespread formation of such structures also during high rates of discharge.

Alluvial Cone Theory. Dietrich (1911); Wunderlich (1929); Flohn (1935).

Almost concurrently with the inheritance theory, an attempt was made to explain the formation of meander valleys with the aid of the impact theory (Sect. 5.2.4.2). This implies that tributaries displaced the main channel and caused it to oscillate, just as in the case of river meanders. As the channel has little chance to move away easily, it would have to begin lateral erosion and start on a bend. The alluvial cone of the incoming tributary is seen as the chief cause, leading to a shifting of the channel line of the main river.

Accordingly, the tributaries should virtually all enter the main channel from the concave side of the bends as clear proof of the respective bend having been caused by this tributary. The opposite also was assumed, however, with the explanation that the lateral cone slightly dammed up the main channel which then worked its path more into the mouth of the tributary causing so to speak an antibend. This theory completely ignored the facts of river hydrology and is only of historical interest.

Rock Meander Theory. Scheu (1909); Vacher (1909); Wagner (1919, 1931); Gradmann (1931); term by Flohn (1935).

During the morphological investigations of ancient shields and tablelands the importance of the underlying rock types soon become recognized. Particularly conspicuous was the change of shape of the meander valley loops resulting from the change in lithological character. The hardness of a rock certainly has a great influence on the development of meander valleys, just as is the case in more or less consolidated materials.

The effect of tectonically induced movements was not considered to be significant and was partly even directly refuted (Wagner). Plain sinusoidal bends and the change-over from "soft" to "hard" rocks were thought to be sufficient. The discussion ran to meander valley-prone rocks (hard limestones, dolomites, sand-stones etc.) and to rocks hardly productive of meander valleys. Vacher even coined the phrase "terrains à méandres", i.e., terrains for meanders. The generally applicable observations in this theory regarding the effects of a change of lithology were devalued by one-sided interpretations and consequently also this concept could not validly establish itself.

Meander Valleys and Tectonics. v. Richthofen (1901); Deecke (1917, 1926 in Maull 1958); Wagner (1919, 1931); Philippson (1924); Schmitthenner (1920); Cole (1930); Masuch (1935); Flohn (1935); Moore (1926); Mahard (1942); Inglis (1949); Ermeling (1951); Leopold et al. (1964); Hunt (1969); Gardner (1975); Schumm (1977 b).

As indicated by the extensive list of authors, long-term movements of the crust are seen as the main factor controlling the formation of meander valleys. Clearcut relations are recognizable, which are most conspicuous in regions underlain by only slightly inclined sediments or in slowly ascending areas.

Strike and dip of the strate as well as cleavage only rarely exert any influence on the formation of meander valleys. In certain areas joints and faults cause the formation of entire valleys or parts thereof. The meander valleys, however, appear to largely ignore jointing sytems. Some exceptions have been reported by Hack and Young (1959).

In the field of tectonics, uplift and/or tilting of the land surface exert the only influence. This statement is supported by field observations and partly by experiments (Hunt 1969; Gardner 1975; Schumm 1977 b).

On the Colorado plateau, Hunt observed that the deeply incised meander valleys occur only upstream of major domal uplifts, i.e., where the gradient became somewhat shallower as a result of the vertical movements along the axis. The uplift apparently occurred at such a slow rate that the river was not deflected, but could cut itself into the bedrock at a uniform rate. Downstream of the dome the gradient is larger, and no meander valleys are developed.

The experiments of Gardner confirmed the principles ruling these processes under the conditions described. It was found in general that an excessively large gradient inhibits the formation of a river as well as of valley meanders or that the river gradually destroys them until a new equilibrium state reflecting the new gradient conditions has been reached.

Conclusions. The theories covering the formation of meander valleys discussed above rely mainly on only one major precondition which, although derived from correct observations, tends to give a rather one-sided view, failing to take into account the multitude of possible shapes.

Even if all questions have not yet been completely resolved, at present a number of correct and reproducible observations and investigations are available. We shall attempt below to explain the phenomenon of the meander valleys from the entirety of geological and climatological forces which also lead to the other channel shapes of natural rivers. By their worldwide distribution in space and time, meander valleys present no particular form other than the normal outcome of the earth's eternal contest between exogene and endogene dynamics.

These important natural phenomena will be set out again as follows:

- Uplift, doming, and/or tilting of portions of the earth's crust are necessary for initiating the processes of incision. Similar effects can be caused by the steepening of a river bed by a heavy supply of bedload material.
- Meanders and meander valleys occur only at certain gradients. If they are excessively steep these forms will not be established or existing forms will be transformed. There is thus conformity in behavior between these two groups.
- Meander valleys are found in solid bedrock as well as in unconsolidated sediments. The assumed smaller size and amplitude of oscillation of meander valleys in unconsolidated materials does not appear to be a universal prinicple.
- A change in bedrock lithology from hard to soft or vice versa, also from solid to unconsolidated, leads to partly rather diverse shapes. This process is, however, not a precondition for the formation or inhibition of meander valleys.
- Meander valleys in unconsolidated sediments are more susceptible to erosion; they are thus more readily preserved in solid rocks.
- Large meander valleys, the rivers of which are not in proper relation to the valley size, were established during periods of higher discharge. There are still meander valleys containing rivers in proper size relation, i.e., the meander valley continues to evolve and is thus "alive".
- Inheritance is not a precondition of formation, but can also be effective when the actual conditions (uplift, gradient) are met.
- There are two main configurations of meander valleys: Those with symmetrical slopes, the entrenched meanders, and those with asymmetrical slopes (concave/convex bank) or ingrown meanders.

From these observations it can be concluded that the sole cause for the formation of meander valleys is the tendency of the river to establish an adjusted gradient (Sect. 5.2.3). The two main disturbing factors which preclude the river from almost ever attaining this gradient are

- *tectonics:* uplift or doming of parts of the crust with or without tilting, rarely tilting alone.
- *climate:* or, better, climate-induced morphology, sharpening of the gradient by excessive supply of bedload material for example during the Quarternary glaciations or generally during times of increased discharge.

The crustal movements, which are usually of longer duration, can naturally also take place during colder periods, leading to the superimposition of these phenomena.

The multitude of shapes results from the possibilities for variations offered by the surface of the earth. The initially rather heterogeneous character of the meander valleys can thus be related to the two factors described, movements of the crust probably being the most prominent.

The causes for the difference between the symmetric and asymmetric sides of meander valleys have not yet been resolved. Since the asymmetric form may be considered as a migrational phenomenon in the horizontal and vertical direction, i.e., the meander valley as such moves downstream to a certain degree, the symmetrical incision gives the impression of topographic stability. The eventual clarification of this question could be achieved by structural investigations.

5.2.4.2   River Meanders

Theories. The origin of river meanders has been the subject of several dozens of theories, most of which, according to Hjulström (1942) deserve the description of being "more metaphysical than physical." Even those theories which have to be taken seriously from a scientific point of view are so numerous that only a limited selection can be presented here. Those interested in more details of the historical development of the theories on meanders can be referred to the compilations of Kaufmann (1929); Hjulström (1942), Baulig (1948); Spengler (1958), Leopold et al. (1964), and Zeller (1967).

The so-called impact theory postulates that accidental external causes, such as a tree fallen into the water, lead to a lateral displacement of the river and the retreat of the opposite bank. Once the flow has deviated from its original path, the initial bend is followed by a second and third one. This concept does have a certain appeal and was supported among others by Davis (1903) and Gravelius (1914). However, it is not able to explain the rather regular amplitudes and arc lengths as the most important characteristics of meanders and was thus abandoned long ago.

The proposal of Exner (1919) was based on similar lines. As a conceptual model he introduced a slightly inclined channel with semi-circular cross-section, in which a sphere introduced outside the center line is rolling downhill. It will never follow a straight course, but instead always deviates to the right or left of this center line and thus describes an oscillating movement. Exner compared the course of this sphere with the movement of water in a sinuous river, the water level of which, due to the centrifugal force, is at one time higher on the left-, the next time on the right-hand bank, and thus forms a standing wave. Model tests in a 50-cm-wide, sand-bottomed flume were carried out to support the analogy between the oscillations of the sphere and the flowing water. In order to obtain the expected oscillations the flow direction at entry of the flume had to be rather steeply inclined. Even then, after only a few lateral deviations the movements were dampened as soon as the energy of the water reflected from the opposite bank had been dissipated. This contradiction between theory and experiment considerably detracted from the merits of the former.

Exner not only attempted to explain meanders as the consequence of standing waves, but also developed a formula for the relationship between the flow velocity and meander wave length l_M, the width of the meander belt b and the water depth h. In the presently used form the flow velocity is expressed as:

$$v = \frac{l_M}{2b} \sqrt{gh} \ . \tag{87}$$

For a number of rivers from central Europe and Northern America this equation has usually yielded excessive velocities.

This concept of the transverse oscillations of the water level as the cause for meanders was also accepted by Hjulström (1942) as a starting point for his own considerations. As an additional criterion he introduced the turbulence of the water, which expresses itself in a dampening and retardation of the oscillation and

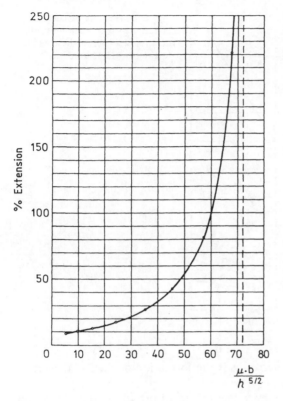

thus as an increase of the wave length. With the coefficient of turbulent friction $\mu(\text{cm}^2\,\text{s}^{-1})$ being constant, the expanded equation for the wave length reads

$$l_M = \frac{2bv}{\sqrt{gh}}\left[1 + \frac{1}{100}f\left(\frac{\mu b}{h^{5/2}}\right)\right] . \tag{88}$$

The values for function f can be derived from Fig. 5.12.

An example from the Mississippi will explain the method of calculation: With Exner's values for the channel geometry, i.e., $h = 10$ m, $b = 30$ km, and $v = 1.3$ m s^{-1}, Eq. (87) results in a wave length of 7875 m. The map, however, suggests rather 13 000 m, i.e., about 65% more than the calculated value. From Fig. 5.12 the necessary correction for the influence of the turbulence is found as $\mu b h^{-5/2} = 54$ and furthermore $\mu = 570$ cm^2 s^{-1}.

The wave will be stretched out when $\mu b h^{-5/2}$ is above 50. Beyond $\mu b h^{-5/2} = 72.1$ the function becomes asymptotic, which is equivalent to an extension of the wave length to infinity. In other words, the influence of the turbulence is so large that no more transverse oscillations are possible and the formation of meanders will cease.

With the limiting value of 72.1 for $\mu b h^{-5/2}$, the maxima of the coefficient of turbulence can be calculated as a function of the water depth and the width of the meander belt. Figure 5.13 illustrates which values for the turbulence must

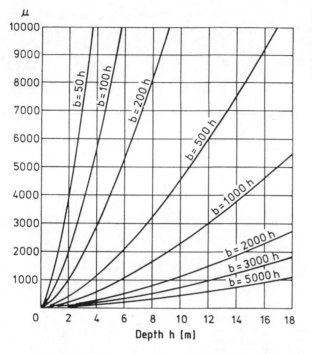

Fig. 5.13. Maxima of coefficient of turbulence in relation to water depth and width of the meander belt. (After Hjulström 1942)

not be exceeded at certain river dimensions if the formation of meanders were to be possible.

As the diagram shows, the turbulence factor is most pronounced at low water depths. In principle, rivers tend to meander strongly when the following conditions are fulfilled:

— great water depth
— low flow velocity
— low turbulence.

The latter two conditions are frequently met in lowland rivers. Rapidly flowing channels with high turbulence, on the other hand, do not form meanders. As the velocity depends on the gradient, there is an unmistakable relationship between the rules presented by Hjulström and the limits for sinuous river courses expressed in Eq. (86) of Leopold and Wolman at least on a qualitive basis. Low water depths are encountered in small creeks and also in larger rivers which silt up by the deposition of sediments such as in the wide channels of a delta. In both cases the conditions for the formation of meanders are not met.

In addition to the hydrodynamic interpretation of the meander problem, Hjulström gives some thought to the properties of the bed material. He comes to the conclusion that easily eroded solids are required. Without transport of material at least at times, river meanders will not form.

Another theory on the origin of meanders was based on Baer's law (Chap. 3.2.1). In every bend a centrifugal force directed to the outside is effective which, due to the irregular vertical distribution of flow velocity, results in a helically twisted flow (Chap. 3.1.3).

   In straight reaches under the influence of the Coriolis force (Chap. 3.2.2) a
similar, albeit weaker, circulation is caused which on the northern hemisphere
leads to a clockwise downstream rotation of the water. The erosion consequently
would have to be stronger on the right than on the left bank. Due to the mass
inertia the circulation attains its maximum value only behind the crest of the
curve, just as does the natural asymmetry of the channel section. This forces the
meander loops to migrate in the direction of flow. No one less than Einstein
(1926) supported this theory and by the sheer weight of his name contributed to
its acceptance.

   As shown in Chapter 3.2.2 the Coriolis force becomes effective in relation to
the force of gravity only in rivers with very small gradients. The rotation of the
earth could thus effect the formation of meanders only in rivers with a very slight
longitudinal gradient. In reality, however, everywhere on earth sinuous river
reaches are developed, the gradients of which are much too large to admit the
Coriolis force as an important contributor to the shaping of their beds. Furthermore, the migration of the meander loops in the northern hemisphere would
always have to be to the right. This does not apply to the larger number of the
sinuous rivers, and so this theory nowadays is only of historic interest.

   Wundt (1941, 1949, 1962) saw the main cause of meander formation in the
contest between erosion and accumulation, as also in other morphologies (Sect.
5.2.7). During a phase of quiescence between these two forces meanders will form
as a natural state of equilibrium. Physically this can be interpreted by the "principle of least coercion". Applied to hydromechanics this implies that a river with
an unimpeded course on a substrate of easily eroded material will shape its bed
so that it offers the least flow resistance. Thus a state of equilibrium is established
between the gravitational energy of the rivers and the resistance of their substrate
"which expresses itself in the temporal and geographic collusion of erosion and
accumulation".

   This concept is supported by numerous field observations such as the imbrica-
tion of bedload particles in river reaches with latent erosion. They present the
maximum resistance to a change of position whereby the flow itself suffers the
least resistance.

   Furthermore Wundt tried to establish dimensional relations between the shape
of meanders and the discharge. The starting point here is the lateral erosion exerted onto the channel wall due to the centrifugal force and the assumption that
the flow velocity is inversely proportional to the length of the circular arc of the
meander.

   The results of this rather extensive derivation imply that the discharges of
comparable channels relate to each other as the squares of the respective arcs of
curvature. The controlling discharge in this case is the mean flood. The data of
Table 16 appear to support this relationship.

   The actual ratios are thus:

$$57 : 3000 : 30000 \approx 0.2^2 : 1.5^2 : 4.5^2 \ .$$

With the aid of further mathematical considerations, Wundt obtained a value of
67° for the center angle of half the meander arc at which the lateral erosion is
highest and the vertical erosion smallest. The observation that natural meanders

Table 16. Radius and mean flood discharge of meandering rivers

River	Mean radius (km)	MHQ (m³/s)
Pegnitz near Nuremberg	0.2	57
Rhine (middle sections)	1.5	3 000
Lower Mississippi	4.5	30 000

usually show larger center angles he ascribed to only part of the total energy of the flowing wave being available for erosive work whereas the remainder is dissipated thermally. Just as according to the second law of thermodynamics the entropy tries to achieve a maximum, the energy exchange of a river is thought to be directed also toward the maximum heat gain. This rather speculative, not very convincing analogism was influenced by the interpretation of a meander as a "rhythmic phenomenon" proposed by Kaufmann (1929). This author saw the origin of meanders in the interaction between water and solids in the common zone of mixing. He wrote: "Thus our 'rhythmic phenomena' can be understood as the expression of a tendency for minimum internal friction in the boundary zone between two media moving along or into each other. Theoretical support is furthermore furnished by the second law of thermodynamics which also characterizes the direction of these natural processes. According to this, all self-initiated processes are irreversible, i.e., the entropy of these changing systems increases and finally they will all end up in a state of equilibrium."

For Kaufmann the investigation of the cause of meandering was less important than the question as to the purpose of nature when allowing a river to attain a sinuous course. He called it in short "a wonder of stabilization" and in this, as in most of this work, he alludes to concepts which appear to be drawn from Schelling's natural philosophy.

In recent times, Leopold and Langbein (1966) and Langbein and Leopold (1966) have treated the formation of meanders as a probabilistic problem. They assumed that meanders are not arbitrary natural phenomena, but rather represent those particular types of channel geometry in which the river has to carry out the least amount of work. Meanders thus have the highest probability rating. A decisive influence on the formation of meanders is not exerted by local irregularities, but rather by accidental factors such as the turbulence interacting between water flow and river bed. It is a natural paradox that such accidental processes result in ordered shapes, just as on the other hand frequently orderly processes may lead to irregular shapes.

For the mathematical description of this concept Leopold and Langbein used the random-walk technique developed by von Schelling. It deals with the task of determining the most probable path covered accidentally by a point at a given length of travel between two locations in a plane. Assuming that both locations are on the positive x-axis of a rectangular system of coordinates, each element, ds, of the initially unknown curve has an inclination, dy/dx, against this axis. The function

$$\varphi = \text{arctg} \; \frac{dy}{dx}$$

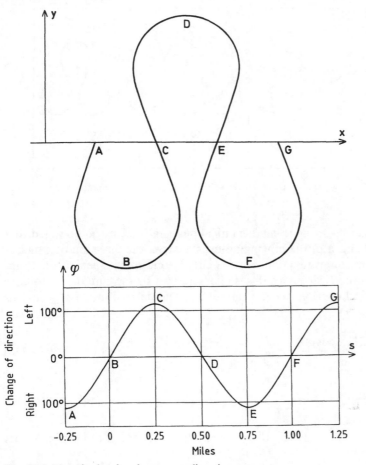

Fig. 5.14. Meander bend and corresponding sine curve

describes the changes of direction of the curve (Fig. 5.14). With the further assumption that the function of the direction $k = d\varphi/ds$ is normally distributed, i.e.,

$$f(k) = C\exp\left[-\frac{k^2}{2\sigma^2}\right]$$

the arc length results as the elliptical integral

$$L = \frac{1}{\sigma}\int\frac{d\varphi}{\sqrt{2(\gamma-\cos\varphi)}}\;, \qquad (89)$$

with the integration constant

$$\gamma = \cos\omega\;,$$

ω representing the largest deflection from the x-ratio (Fig. 5.15).

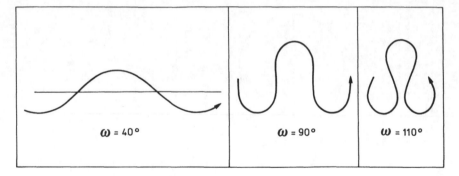

Fig. 5.15. Theoretical sine-generated curves. (After v. Schelling, from Langbein and Leopold 1966)

Without further resorting to mathematical deliberations, it can be stated that the most probable path of the moving point is a curve, the direction of which is described by a sinusoidal function. Such a curve is called sine-generated. Compared to other geometrical curves this is characterized by the sum of the squares of the directional changes k being a minimum. The deflection from the x-axis amounts to about

$$\varphi = \omega \sin \frac{\sigma \sqrt{2(1-\cos\omega)}}{\omega} s , \qquad (90\,\text{a})$$

or

$$\varphi = \omega \sin \frac{s}{M} \cdot 2\pi . \qquad (90\,\text{b})$$

In this M represents the total length of a wave. Within the range of possible values for ω there will be

$$\omega / \sqrt{2(1-\cos\omega)} \approx 1.05$$

and thus, according to Eqs. (90a) and (90b)

$$M \approx \frac{6.6}{\sigma} ,$$

i.e., inverse proportional to the standard deviation.

A sine-generated curve can be considered as the basic shape of a river meander. This implies that the direction of each point on the river axis on the thalweg in relation to the direction of the valley represents a sine function of the river length (Fig. 5.14). At the crest (B, D, and F) both directions coincide and the angle of deflection is zero. At the points of inflection (A, C, E, G) the angle of deflection attains its maximum. The comparison of loops of typical meandering rivers, but also of meanders generated in laboratory flumes, has confirmed the analogy between the mathematical calculations and the natural morphologies. Figure 5.16 presents the example of a loop of the Mississippi together with the corresponding directional function.

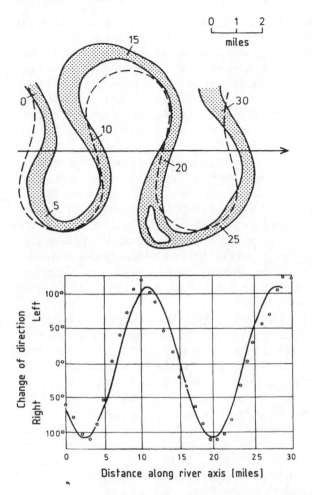

Fig. 5.16. Mississippi at Grenville with sine curve meander. (After Leopold and Langbein 1966)

Another property of the sine-generated curve is that in the mechanical model it requires the least effort in deformation while bending. A small example will be given as an explanation:

A watch coil spring fixed at two points under the influence of forces and moments will attain that particular shape which requires least effort in deformation, irrespective of how the spring is being bent. The effective effort in each length element of the spring will be proportionate to the angle of curvature of this element. If this conceptual model is transferred to river meanders, they have to be visualized as shapes for the formation of which nature has to expend a minimum of work. Wundt's "principle of least coercion" resurfaces here under a different point of view.

As the irregular meander shapes are frequently difficult to simulate by mathematical curves, the approach of Leopold and Langbein was developed further in a more visual-comparative method by Ferguson (1977). He proposed checking the elements of meander geometry on as many rivers as possible, in order to obtain statistical answers from so-called meander spectra.

In view of the many theories on the origin of meanders, one could gain the impression that there are no more open questions. In reality, however, nature continues to generate new shapes, which cannot be explained satisfactorily on the basis of our present knowledge of the laws of physics. Consequently, empirical rules tend to be applied which have been derived from a large number of observations on all continents. It is assumed that sinuous channels are stable despite their migration − actually a contradiction in itself − and that their geometric properties are in well-defined relation to each other, similarly to the implications of the regime theory for straight channels.

In addition to field observations, flume experiments are used for the verification of certain aspects of meander geometry. At the start it must be clarified whether the results may be transferred to the natural situation according to Froude's law of similarity. According to Zeller (1967), the reproducibility appears to be generally assured, although only few data are available. Model tests have the advantage that they can be carried out under uniform and exactly defined conditions and are not subject to the disturbances unavoidable in nature. One disadvantage is the sometimes rather coarse derivation of the controlling parameters and of their interaction on a few measurable factors. Model tests thus will never be able to replace field observations completely.

At first the range has to be defined within which meanders can be formed at all. As mentioned already for braided rivers (Sect. 5.2.3), the boundary between them and the sinuous rivers is defined by Eq. (86) and Fig. 5.11, respectively. Furthermore, the Froude number appears to be an important criterion. It is assumed that meanders occur preferentially in the range $Fr = 0.2-0.5$, at times also up to 0.7. At larger Froude numbers, especially above 1.0, occasionally so-called pseudo-meanders are developed, the channel shape of which has some similarity with "true" meanders. They are, however, caused by different hydraulic conditions, i.e., standing waves in an initially straight channel with dune fronts vertical to the axis. When these dune fronts start at super-critical flow to shift into an acute angle with the banks, oscillating erosion rills are formed which give rise to transposed bank erosion and thus to apparent meanders. This phenomenon will not be discussed further.

The large quantity of respective investigations prohibits their separate discussion. Therefore, only a few of the best-known formulas for the basic parameters of meander loops will be presented here in accordance with Zeller. For the designations of Fig. 5.2 there is:

1. meander length l_M

$$l_M = K_1 \cdot Q^{c_1} \tag{91a}$$

$$l_M = K_2 \cdot A_E^{c_2} \tag{91b}$$

$$l_M = K_3 \cdot B^{c_3} \tag{91c}$$

2. meander amplitude b_M

$$b_M = K_4 \cdot Q^{c_4} \tag{92a}$$

$$b_M = K_5 \cdot B^{c_5} \tag{92b}$$

$$b_M = K_6 \cdot l_M^{c_6} \tag{92c}$$

Table 17. Coefficients and exponents of Eqs. (91) – (93) in metric system

Reference/ Parameters	Leopold and Wolman	Inglis	Shaw	Dury	Altunin	Hack
K_1	64	65.5	49.8	54	–	–
K_2	–	–	–	9.5 – 16.1	–	11.3
K_3	6.5 – 15	6.6	–	–	12 – 15	–
K_4	–	10.4	–	–	–	–
K_5	2.7	18.6	–	–	–	–
K_6	–	1.7	–	–	–	–
K_7	2.3	–	–	–	–	–
K_8	0.21	–	–	–	–	–
c_1	0.5	0.5	0.5	0.5	–	–
c_2	–	–	–	0.5 – 0.66	–	0.5
c_3	1.01	0.99	–	–	1.0	–
c_4	–	0.54	–	–	–	–
c_5	1.1	0.99	–	–	–	–
c_6	–	1.06	–	–	–	–
c_7	1.0	–	–	–	–	–
c_8	1.0	–	–	–	–	–
Type of Q	Bank-full stage	Probability of annual recurrence = 1				

3. radius of curvature r

$$r = K_7 \cdot B^{c_7} \tag{93a}$$

$$r = K_8 \cdot l_M^{c_8} \tag{93b}$$

The coefficients and exponents of Eqs. (91) – (93) can be found in Table 17.

The best-defined rule is given by Eq. (91 a). Equation (91 b), which correlates the arc length with the size of the drainage area, incorporates some uncertainty, as each river basin has certain special features which are not directly applicable everywhere else. All other equations have to be treated with care.

In accordance with their structure, some of the equations are dependent on each other. Their division results in the conditions

$$\frac{K_1}{K_4} \approx \frac{K_3}{K_5} \quad \text{and} \quad \frac{K_8}{K_6} \approx \frac{K_7}{K_5} \; .$$

As far as the coefficients are variable, it should be ensured on selecting then that these equations are fulfilled at least approximately.

The rules discussed so far have the disadvantage that they neglect parameters such as gradient and bedload grain size and that they have pure dimensions. There is no doubt that a comprehensive meander theory still has to be established on a wider basis.

Migration. Up to now meanders have been treated as stationary features. In reality they widen their loops at various rates and move downstream. The latter process is described as migration. The two phenomena can be developed separately, but usually they are interlinked. Migration, during which the original shapes are

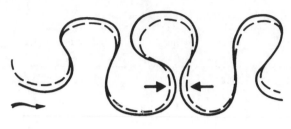

Fig. 5.17. Widening, migration, and cut-off of meanders. (After Zeller 1967)

usually more or less preserved, progresses at different rates. Hjulström described Swedish rivers which barely changed their position and thus have to be considered as inactive. American rivers migrate at velocities of up to $200 \, \text{m a}^{-1}$. For the Mississippi at Rosedale, for example, during 1881 – 1913, rates of $192 \, \text{m a}^{-1}$ were measured. The erosional resistance of the bank appears to figure prominently in the velocity of migration.

When the meander loops approach each other due to their lateral erosion, a cut-off is established at the contacts, which leads to the strangulation of part of the old river bed (Fig. 5.17) and to the establishment of so-called oxbow lakes.

Although cut-offs are not very frequent, it is to them that active meandering rivers owe a considerable part of their dynamics. Each cut-off shortens the course and thus leads to an increase of the gradient in the area thus affected. This higher gradient leads to an increase in the energy potential of the flowing wave and due to the increased erosion accelerates the widening of the loops. The shortened river tries to achieve a new quasi-stable stage, until the process is repeated at the next cut-off. It should be pointed out that the river bank vegetation frequently exerts a profound influence on the process of bed formation, especially when fallen trees deflect the flow or offer considerable resistance to erosion with their roots. These influences, however, are only variations in the physically controlled meander morphology, and play no part in the actual origin of these structures.

5.3 Longitudinal Section

5.3.1 Channel Gradient

The longitudinal profile of a river is one of the most important morphological elements to mark the earth's surface. The channel courses define the base levels of erosion in all humid climatic regions. Their development controls in a certain sense the morphological character of a landscape, as in all humid and semi-humid regions rill-like erosion predominates. Even in arid regions where sheet erosion is much more predominant, linear developments cannot be excluded (v. Wissmann 1951; Louis 1961; Wilhelmy 1972).

When looking over a river course in the field, the longitudinal profile is less obvious than the cross-section, but it still exerts stronger formative influences. Although shape and course of this morphological element are no random features, their mathematical definition, although frequently postulated, meets with great difficulties. The longitudinal profile of a river represents its gradient or, in other words, the inclination of the bottom at each given point as a reaction to tectonic, lithological, and climatic factors. The question as to the origin of a certain type of profile has been debated for some time, as the interaction of the forces exerted, especially during erosion, was never clearly defined in theory, nor has the joint consideration of geological factors and hydrological laws ever been satisfactorily set out.

The general course of a longitudinal profile is quite clear. The gradient in the upper reaches is steep, it decreases gradually downstream, and is very low in the delta reach, dwindling to zero at sea level, the base level for the erosion. This profile results in a concave curve which was early investigated with the aim of finding in it a geometric, i.e., mathematically defined shape or, even better, a principle of formation. Engineers and scientists took different ways to do so. However, a strictly defined geometric curve is never developed. Already Penck (1894) refuted several of such theories from a scientific point of view. This does not, however, negate the possibility that over partial reaches the longitudinal profile can indeed be mathematically defined (Sect. 5.3.2).

Dividing a river into its upper, middle, and lower reaches has become well established. Erosion is predominant in the upper reaches, transportation in the middle, and sedimentary accumulation in the lower. This concept was refined by a further subdivision of the middle reach into a braided section in which accumulation does occur and a less inclined meander reach in which approximate equilibrium has been established (e.g., Wundt 1953). This subdivision cannot be rejected out of hand, although it is highly schematic. The terms were introduced by A. Heim (1878 in Karl 1970) in order to describe the erosive and accumulative morphology of mountain torrents. However, they could only be applied to an "ideal" river coming from a mountain range, flowing through forelands and several ranges of hills and eventually, after passing through a wide plain, discharging into the sea.

The same system is preferred by Schumm (1977b), who subdivided river courses into three zones:
- zone 1: production, drainage basin
- zone 2: transfer
- zone 3: deposition.

In this, zone 1 corresponds to the upper, zone 2 to the middle reaches, zone 3 to the area of the mouth, without initially considering the size of the total drainage area.

Schumm was aware of this idealization, erosion and deposition also being active in the other zones. He wished to express that in each zone essentially only one morphogenic force was active. The illustration presented by him, however, is strongly reminiscent of the configuration of the drainage area of an ideal mountain torrent and thus suggests stronger morphological differences. The drainage area of large rivers, especially in lowlands, would be extended in favor of the middle and lower reaches, resulting in a difficult and impractical scheme of subdivision. For clarification these stages are shown in Fig. 5.18. The lowland river (bottom) would be several times longer than the mountain river (top) and would furthermore consist of only the lower reaches.

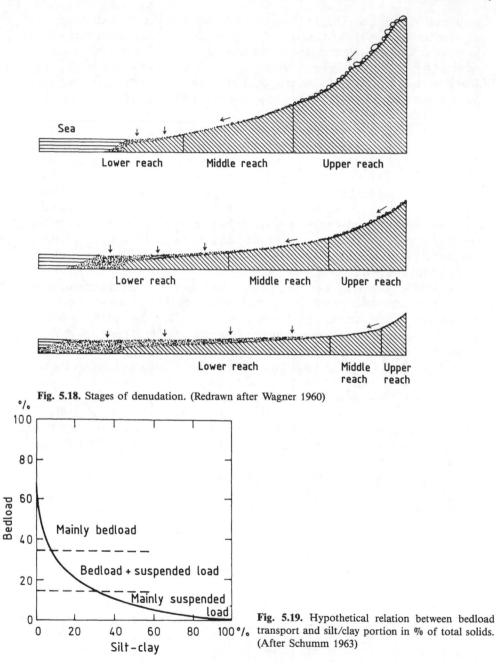

Fig. 5.18. Stages of denudation. (Redrawn after Wagner 1960)

Fig. 5.19. Hypothetical relation between bedload transport and silt/clay portion in % of total solids. (After Schumm 1963)

Preferable to this would be the classification of channels according to morphometric criteria on the one hand, and to the grain size distribution on the other. Schumm (1963) also proposed a classification for the evolution of the longitudinal profile, He plotted graphically amongst other factors the proportion of bedload in the total sediment load against the silt/clay portion and obtained a hypothetical curve of various channel types which is similar in shape to that of the up-per, middle, and lower reaches, being based, however, only on sediment yield and grain size (Fig. 5.19). From this ratio and the channel shape he developed nine different types of channels. This introduced a genetic component which covers one sole aspect, i.e., purely geological factors. Schumm wanted to make mapping in poorly accessible regions easier.

In principle the river tries to achieve a gradient at which under the actual con-ditions of discharge it can transport just the amount of bedload supplied. The gradient is thus controlled by the supply of bedload and its transport capacity (cf. Chap. 4.2.7). If this is too large in relation to the bedload being brought in, the river aims for a reduced gradient by lowering its bed. Such disturbances can be caused by increases in discharge, narrowing of the bed, or by a reduction of the bedload supply. The opposite factors lead to a steepening of the gradient, and to the river locally accumulating solids and building a stretched alluvial cone. Far-ther upstream where the longitudinal profile can be more inclined, redeposition reaches with an equilibrated bedload budget are frequently found. The longitudi-nal section in every case represents the so-called equilibrium gradient.

Under the aspects of tectonics, the lithology of the substrate, and the climate, which jointly control the gradient, bedload, and discharge, the longitudinal pro-file of a river acquires its own configuration. As an example a simplified profile of the Rhine is presented (Fig. 5.20). Three times the river attempts to establish its concave (or convex-downward) equilibrium curve. The Alpine Rhine loses its gradient and bedload on entering Lake Constance. The Upper Rhine loses part

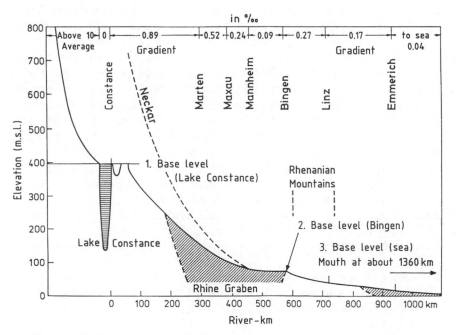

Fig. 5.20. Longitudinal profile of the Rhine

of its gradient in the Rhine Graben, the Rhenanian Mountains acting as a weir. Its final erosive base level is the North Sea.

This break in gradient is usually referred to as base level, a term that does not have to be taken literally. The phenomenon of several base levels being present along the course of a river is widely developed. The more irregular the relief and thus the geological substrate, the more "steps" will be developed. Such longitudinal profiles were even called stepped valleys by Schaffernak (1950). This is rather confusing and not to the point when comparing it to the real stepped valleys caused by glacial downcutting in the main valley in relation to the side valleys. Ahead of each obstacle occurs a reduction of the gradient, the obstacle itself then being overcome by an increase in gradient.

The overall disposition of the longitudinal river profile follows the distribution of the global climatic zones. As mentioned above, in humid zones rill erosion predominates, not in the least because of the general vegetational cover. The gradients here are usually steeper than in arid regions.

Nevertheless, the resulting curves here are also almost exclusively concave (Mortensen 1942; v. Wissmann 1951; Louis 1961). The upward-convex reaches which had earlier been postulated for arid regions could not be confirmed by check surveys. Wherever they are developed, they are only of short extent and tied to obstacles in the field. The same situation has been reported from the semi-arid southwest of the United States (Leopold et al. 1964; Lee and Henson 1977).

5.3.2 Equilibrium Gradient

Wherever the free play of forces has established a state of dynamic equilibrium between the shear stress of the flowing water and the resistance of the mobile bed of the channel, an equilibrium gradient or shape is developed. The process as such has been known for quite some time, but it was Sternberg (1875) who first brought the configuration of the longitudinal profile into an analytical form. His theory was based on the following concept: When a bedload particle moves downstream, along a plane bottom a state of equilibrium between the increase of force and frictional resistance could only be maintained for an undiminished grain size. However, in relation to the energy produced, the particle is subjected to attrition. If the state of equilibrium is to continue, the energy surplus resulting from the reduction of the particle weight must be dissipated by an increase in the length of path covered, in other words the gradient is reduced concomitantly with the reduction in particle size. If the equilibrium gradient is larger than required by the frictional resistance, the river will incise itself by dint of its surplus of energy. Under the opposite conditions the bed will be aggraded. These processes continue overall until the gradient everywhere has been adjusted to the frictional resistance of the local bedload situation.

The rather cumbersome calculations of Sternberg have been refined by and adjusted to practical requirements by Putzinger (1919, 1927). The basic equation for the bottom slope downstream over the distance x from the starting point is

$$J_x = J_0 \cdot e^{-vx} \ . \tag{94}$$

This equation is structured as Sternberg's law of attrition [Eq. (43)] and expresses the relationship between the reduction in gradient and particle size.

The so-called gradient reduction number, which must not be confounded with Sternberg's coefficient of attrition, can be derived from the comparison of nearly stable reference reaches at the distance l. In logarithmic form Eq. (94) reads

$$v = \frac{\ln J_0 - \ln J_x}{l} \ . \tag{95}$$

In reality v is actually not a constant applicable to the complete course of a river. The lateral input of bedload, geological factors, hardness and resistance of the underlying rock to abrasion exert an influence on the gradient reduction number and, especially in the upper reaches, can lead to abrupt changes. Putzinger was aware of these facts and modified his basic Eq. (94) accordingly by corrective factors which may be neglected here.

As there is $J = dy/dx$, Eq. (92) after separation of the variables and integration results in an equation for the longitudinal profile

$$y = \frac{J_0}{v} e^{-vx} + C \ . \tag{96a}$$

The curve is asymptotic to the x-axis at the distance $y - J_0/v$. As the choice of the coordinate system is free, one usually sets $c = 0$ and the x-axis as the asymptote to the curve, the equation of which now reads

$$y = \frac{J_0}{v} e^{-vx} \ . \tag{96b}$$

Setting $x = 0$, there is $y_0 = J_0/v$, allowing a transformation to

$$y = y_0 e^{-vx} \ . \tag{96c}$$

The height y relates to the x-axis as the base line and thus bears no relation to the height above sea level. If this is to be included, the equilibrium may be tied to the natural profile at any locality.

The application of the equation to the Danube and the Inn will illustrate their use. The origin of the coordinates for the Danube with the x-axis being directed downstream is taken at a position about 10 km above Ulm. As reference reaches the following sections will be taken

- Berg-Ulm   0.89‰ average gradient
- Kelheim-Regensburg   0.325‰ average gradient

At about 220 km distance between the reference sections one finds

$$v = \frac{\ln 0.00089 - \ln 0.000325}{220} = 0.0045 \text{ km}^{-1}$$

and

$$y_0 = 0.00089 : 0.0045 = 0.194 \text{ km} \ .$$

The ordinates of the line of equilibrium are

$$y = 194 \cdot e^{-0.0045x} \ .$$

The natural longitudinal profile of the Danube at the mouth of the Isar and especially of the Inn is subjected to considerable increases in gradient. The lower reference reach thus has to be selected sufficiently far upstream of these points of disturbance. Down to Passau the natural and the calculated profile of the Danube do not deviate much from each other. Below Passau there is, however, no agreement at all.

For the morphogenetically less mature Inn an equilibrium curve which would be valid over longer reaches cannot be calculated. When comparing two reasonably well adjusted sample reaches at a distance of 140 km, i.e.,

Innsbruck   1.60‰ average gradient
Wasserburg   0.82‰ average gradient,

one finds

$$v = \frac{\ln 0.0016 - \ln 0.00082}{140} = 0.00477 \text{ km}^{-1}$$

and

$$y_0 = 0.0016 : 0.00477 = 0.335 \text{ km} \ .$$

The ordinates of the line of equilibrium are

$$y = 335 \cdot e^{-0.00477x} \ .$$

If the calculated curve is tied at Passau to the profile of the Danube, then there are considerable deviations from the natural longitudinal profile in the upstream direction, especially beyond Innsbruck. Below Passau the calculated line of equilibrium of the Inn initially deviates little from that of the Danube. This situation can be interpreted by the morphological picture of the Danube in this section being controlled mainly by the Inn. Only below the mouths of the bedload-laden Traun und Enns rivers do both curves deviate markedly from each other, as these two tributaries supply the Danube with fresh coarse bedload, so that the mean grain size can be considered as being uniform over a prolonged reach. A constant grain size, according to Sternberg, is the precondition for an unchanged gradient. The longitudinal section of the Danube down to Bratislava is in fact almost straight and stretched. In Fig. 5.21 the calculated profiles are compared to the natural ones developed prior to the construction of the hydro-power plants.

Criticism has been voiced from various quarters of the application of Sternberg's concept for the calculation of the longitudinal section of a river. Already in 1883 v. Salis, as cited in Putzinger (1954/55), remarked: "Thus a formula which is to express the formation of the longitudinal profile will be correct only if it properly expresses the sum of the effective influences, and not if this is the case for only some of these factors."

Methods of calculation are generally designed for abstraction and will never be able to express in detail the intricate natural processes. Thus care must always be taken that any theoretical assumptions at least tend to follow the laws of nature. It would be wrong to use a morphogenetically immature reach for the calculation of the gradient reduction number.

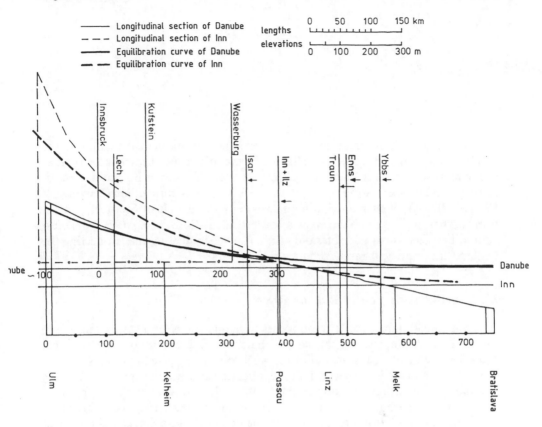

Fig. 5.21. Sketch of longitudinal profile of Danube and Inn

Over and above v. Salis' criticism, Forchheimer (1930) noted that in his considerations Sternberg assumed a stationary river bottom, whereas in reality the transport of bedload tends to destroy this static state of the bed. These critical remarks cannot diminish Sternberg's merits in having developed the first practical method for the calculation of the equilibrium gradient.

An improved version of the formula, which takes the transport of bedload into account, was developed by Meyer-Peter and Lichtenhahn (1963). It is based on a rectangular channel having constant width and discharge. The expression of the equilibrium gradient is transformed with the aid of Strickler's law of discharge [Eq. (8)], Sternberg's law of attrition [Eq. (43)], and Meyer-Peter's law of bedload transport [Eq. (67)] into the following form:

$$\frac{J_x}{J_0} = e^{-3/7c} \left[\frac{1 + Ce^{-3/7cx}}{1 + C} \right]^{10/7}, \tag{97}$$

with the abbreviations

$$C = \frac{\varepsilon g_0''^{2/3}}{\beta \cdot d_0}$$

$$\beta = 0.047 \cdot \gamma_s''$$

$$\varepsilon = 0.25 \left(\frac{\gamma_w}{g}\right)^{1/3}$$

If J_0 of the sample reach is known, the formula permits the calculation of the equilibrium gradient for any other reach, provided that no disturbances such as rocky narrows, strong tributaries, etc. intervene. In the case of no bedload transport, C will disappear and Eq. (97) with $v = 3/7c$ is transformed into Putzinger's Eq. (94). The gradient reduction number is thus about half of Sternberg's attrition coefficient. As can be shown by a few test calculations, in Eq. (97) the expression in brackets usually deviates only little from 1, indicating that neglecting the bedload transport will rarely produce great errors.

5.3.3 Self-Stabilization of Eroding Rivers

A stable longitudinal profile presupposes a balance between bedload input and output in a certain river reach, the bed material being continuously exchanged. However, a state of "quasi"-equilibrium can also be established in the case of "latent" erosion. This is a special type, representing the final stage of an eroding channel. Its formation requires the ability of the river to form an armor-"plating" of the bed layer leading to a gradual increase of the bed resistance to attack by flow during degradation until, despite the bedload deficit, a nearly stable state of the bottom has been regained. In this case the resistance of the coarsest particles of the bed material remaining as a cover layer or paving after the selective removal of the fines is of prime importance.

For the description of the state of equilibrium it is advisable not to start from the shear stress on the channel bottom, but from the resistance gradient, as this factor implicitly includes the controlling turbulence factors. According to Schöberl (1981), this is a function of the relative roughness ε, the Froude number Fr, and of the grain structure coefficient a which expresses the influence of the composition of the bed material on the formation of the covering layer, or in short

$$J = f(\varepsilon, Fr, a) .$$

Model tests resulted for the grain structure coefficient a in the empirical equation

$$\frac{1}{a} = 0.878 \left(\frac{d_{90}}{d_{50}}\right)^{-1/3} \left(\frac{d_{mGS}}{d_{50}}\right)^{-1/2} ,$$

in which

d_{mGS} = determining grain diameter of the bottom layer

d_{50}, d_{90} = grain size corresponding to 50th and 90th percentile of bottom layer mixture.

As found from comparative calculations, the ratios for the bottom layer distributions in Alpine rivers are usually in the range of

$$2.5 \leq \frac{d_{90}}{d_{50}} \leq 5 \quad \text{and} \quad 1.1 \leq \frac{d_{mGS}}{d_{50}} \leq 1.7 \ .$$

This corresponds to grain structure coefficients of

$$1.6 < a < 2.5 \ .$$

The same experiments further resulted in resistance gradients of

$$J = \frac{1}{a} \left(\frac{\varepsilon}{Fr} \right)^2 \ .$$

Using Strickler's equation converted to the meter-wide strip

$$q = k_s \cdot h^{5/3} \cdot J^{1/2} \ ,$$

the relative roughness

$$\varepsilon = \frac{d_{mDS}}{h}$$

and the Froude number

$$Fr = \frac{q}{h^{3/2} g^{1/2}}$$

the equation for the resistance gradient in its more practicable form reads as

$$J = \left(\frac{k}{q^{7/5} k_s^{3/5}} \right)^{10/13} \ , \tag{98}$$

with

$$K = 8.61 \ \frac{d_{mDS}^2}{(d_{90}/d_{50})^{1/3} (d_{mGS}/d_{50})^{1/2}} \ , \tag{99}$$

d_{mDS} being the determining diameter of the bottom cover.

With Eqs. (98) and (99) the final stable gradient at fully developed bottom armoring can be calculated. The method has been confirmed experimentally for gradients between 1.4‰ and 2%. The reliability of these results depends essentially on the quality of the initial data. It is therefore recommended not only to use average values, but rather to consider also the range of the grain size analyses.

5.4   Cross-Section

As outlined in Section 5.3, the longitudinal profile of rivers can be considered as the base level of the erosion. In every phase of development and in each corresponding portion of its course, a river will also show the appropriate cross-section. Just as the longitudinal profile of a river need not be identical with that of its valley, above a certain size the cross-sections of rivers will bear no relation to those of their valleys. There are, nevertheless, direct and indirect relationships between river and valley, especially as far as the cross-sections are concerned (Wittmann 1955).

5.4.1 Shapes of Valleys

Valleys and their shapes have been frequent subjects of geomorphological research. According to the control mechanisms enumerated in Section 5.1, a river can be considered as a "regulator of valley formation, although it ... can directly shape only its bed and the flood plain. The river influences on the formation of the valley slopes are usually of a more indirect nature ..." (Maull 1958). Rill erosion completely reshapes the slopes as far as the vegetation will allow and as far as the river itself does not attack them by undercutting. A valley thus is the result of the interaction of river erosion and general corrosion, the upper slopes no longer being shaped by the river.

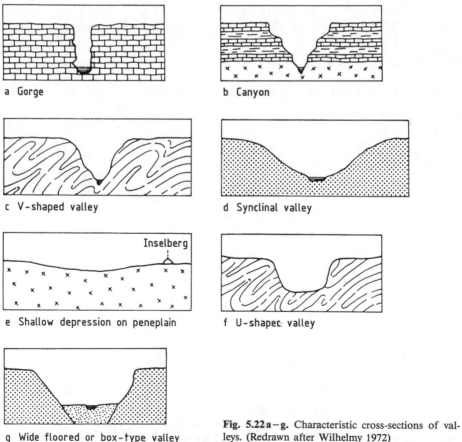

a Gorge

b Canyon

c V-shaped valley

d Synclinal valley

e Shallow depression on peneplain

f U-shaped valley

g Wide floored or box-type valley

Fig. 5.22 a–g. Characteristic cross-sections of valleys. (Redrawn after Wilhelmy 1972)

For a number of valleys the cross-section of the valley is equal to the flow section on the small and large scale. This is, for example, the case for all gorge reaches which are controlled by stratification and hardness of the respective bedrock succession. Such reaches, however, are only short. Allowing some leeway, certain canyons can also be included here, although the lateral slopes can attain considerable dimensions. The river remains the dominating influence, but will soon develop its own individual life. The transition from the identical profile of valley and river to a river in a valley, as seen in profile, is diffuse and can repeat itself depending on the change of lithology and the tectonic situation. Based on lithology and climate, several types of valleys can be distinguished. Following Wilhelmy (1972), the most important ones are presented in Fig. 5.22. It is impossible to discuss here in detail the various relevant investigations.

Direct or at least close relationship to the river is developed only for the gorge, the canyon, or the V-shaped valley, lateral erosion gradually becoming predominant. In shallow valleys sheet erosion will be so strong that the definition of the term valley will become problematic. Within central European mountain ranges − and even more outside of them − terraced valleys are frequently developed, whose investigation is an important chapter in river history. The present rivers are virtually lost in these wide valleys, but they still keep to the base level of erosion.

The erosive activity of the river results in characteristic asymmetries of the slope as erosion proceeds selectively, i.e., the softer rock formations are preferentially removed. The flow tends to spare harder rock sections, although there are also exceptions such as in the case of the Rhine Cataract at Schaffhausen. Pronounced assymmetries originate during the formation of meander valleys (Sect. 5.2.4.1), but also through solifluction because of differential insolation of the valley slopes. Comprehensive compilations on the formation of valleys were presented, for example, by Maull (1958), Wagner (1960), Louis (1961), Weber (1967), and Wilhelmy (1972).

5.4.2 Cross-Section on Rivers

These configurations are the direct expression of the controlling forces of erosion and transportation. A few characteristic cross-sections of rivers are presented in Fig. 5.23. In a gorge the river occupies the entire width of the valley, the bed is rocky and covered by large boulders and a bottom layer of bedload is frequently absent. The bedload coming in is rapidly swept away, the degree of comminution apparently being variable, depending on the roughness of the bed. As investigations by the Bavarian Landesstelle für Gewässerkunde have shown, the calcareous bedload of, e.g., the Ostrach near Sonthofen (upper Allgäu) is comminuted in the Eisenbreche gorge less than had been originally expected.

Depending on the gradient and the lateral supply of bedload, in gorges there is, over longer reaches, already a closed cover of bedload which is still rather coarse and partly accompanied by large boulders. The characteristic feature of all bedload-bearing rivers is first displayed here: a continuously, sometimes rapidly changing river bed with variable cross-sectional shapes.

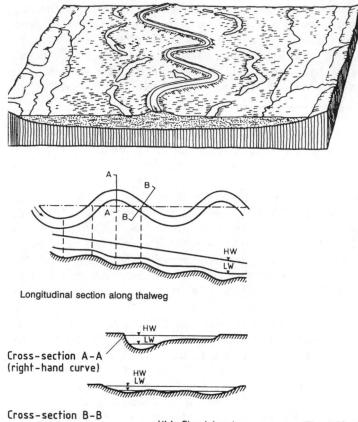

Longitudinal section along thalweg

Cross-section A-A
(right-hand curve)

Cross-section B-B
(transition reach)

HW: Flood level
LW: Low-water level

Fig. 5.23. Characteristic cross-sections of rivers

The movement of water in sinuous river reaches, i.e., in meander loops, leads to a characteristic change in the cross-sections. In the crossings or riffle portion, the river is wider and less deep, whereas in the pool sections of the curves the river is correspondingly deeper and narrower, and furthermore asymmetric (concave/convex banks). This principle also applies to meander valleys.

Zeller (1967) pointed out that sinuous channels are usually wider than straight ones. Fargue (1868) was already of the opinion that the crossings should be narrower than the bends. However, he apparently considered in this respect the already regulated river which serves as a channel of navigation and thus exhibits a shape that does not reflect the naturally established situation. Changes in the width of the cross-section concurrently result in continuous changes in the longitudinal gradient of the river (cf. Sect. 5.5).

A characteristic of many lowland rivers is the elevated position of their course between natural dykes in the form of "levees". Whereas in erosion areas there are a larger number of valley shapes, the levees represent the only specific feature of sediment-laden rivers in regions of accumulation (Wilhelmy 1972). The suspended load settles when the flow decreases or because of the retarding effect of the bank vegetation during floods, leading to a gradual upward growth of the bed and the banks. The levees usually show an asymmetrical section, the coarser material being deposited more within the channel and the finer particles somewhat farther away. The levees have relatively steep banks against the river, while there is a gradual falling away into the surrounding flood plains which frequently lie at a somewhat lower elevation. In this way the back swamps of the Mississippi or real lakes such as its bayous and the varseas of the Amazon basin (Wilhelmy 1958) have been formed.

The water level of the river thus rises above the surrounding plains. During strong floods there is always the danger of the dams breaking, leading to catastrophic inundations and relocations of the course, acquiring huge dimensions such as on the Mississippi, Hwang-ho, and Po River. As, for example, the course of the Po has been narrowed over long stretches by the river regulation, the masses of sediment had to be deposited in a narrower bed, unless they were carried off seaward by the accelerated discharge. As a consequence, strong deposition occurred which forced a continuous artificial increase of the embankments. At the same time the danger of floods increased considerably (cf. Chap. 7.1.1).

5.5 Fargue's Laws

Prior to the advent of the railway age, rivers were the main traffic arteries. No wonder that the boatmen always paid great attention to the shape of the river bed and the navigation channel. Care had to be taken especially to assess the crossing properly, the shallow between two bends, and to pass it without danger to the vessel and its crew. With the start of river regulation as an engineering science – on a larger scale since the beginning of the 19th century – the planners found themselves confronted with the task of typing to comply with the requests of the shipping trade by the proper selection of the various elements of alignment. Whereas initially they relied mainly on direct inspection and experience, gradually the need arose to investigate in detail the laws of channel geometry and to unravel the relations between the individual parameters. The French engineer Fargue (1868) can claim the merit of being the first to have formulated a number of rules, from observations on the Garonne, which have become known in literature as Fargue's laws.

Fargue opened his treatise with the following questions: "Is there a relationship between the channel pattern of a navigable river with a movable bed and the longitudinal profile of its thalweg? What is this relationship? Which shape is best suited to fulfill the interests of the shipping trade?"

Fargue believed he had found the answers from the investigation of the 22-km-long stretch of the Garonne between the villages of Gironde and Barsac somewhat upstream from Bordeaux. In his own words, "observation of facts and logical deduction" describe his approach. He subdivided the stretch investigated into 17 sections which are separated from each other by "points d'inflexion" at alternating curvature and by "points de surflexion" at uniform curvature. For each of the 17 sections of curvature the most important geometric elements such as length, degree of curvature, tangential angle, depth etc. were tabulated and compared with each other. Fargue found that between the configuration of the bed at mean water level and the depth of the navigable channel certain relationships exist which can be combined into the following six empirical rules:

1. The pool and the riffles are displaced downstream in relation to the crest of the arc and/or the point of inflection (law of deviation);
2. the curvature of the arc crest controls the depth of the pool (pool law);
3. with regard to the greatest as well as to the mean depth the arc should be neither too short nor too long;
4. the external angle of the tangent to the curve, divided by the length, controls the mean channel depth (angular law);
5. the longitudinal profile of the thalweg exhibits regularity only inasmuch as the curvature changes continuously and in stepwise fashion; each sudden change of curvature results in a sudden reduction of depth (law of continuity);
6. when the curvature changes in uniform fashion, the inclination of the tangent to the curve controls the bed slope along the thalweg (law of river bed gradient).

In the second part of his treatise, Fargue presented the mathematical expression of his concept. He started from purely geometric considerations, without taking into account the problems of sediment transport which later were to so greatly activate the science of river morphology. He checked, for example, the suitability of a clothoid as an element of design of an arc. The most important partial results of his study were a number of functions illustrating the relationship between the water depth and the degree of curvature of the arcs. A radius of curvature of 1 km according to this correlates with a water depth of 4.8 m in the pool section according to the second of the above laws. He was convinced that with the methods presented by him it should be possible to attach a rational shape to any river, i.e., to bring the curvatures into logical agreement with the longitudinal profile most suitable for navigation. Fargue postulated a general applicability for his equations and graphical presentations.

In their application to other rivers only the coefficients might have to be selected differently, depending on gradient, channel width, discharge, and properties of the bed layer.

A third chapter was dedicated to the theory of streamlines. Amongst other aspects it deals with the relocation and stability of the thalweg.

In a further study Fargue (1882) discussed the width of the bed of the Garonne at mean water level. The central part of this publication is the relationship between curvature and the width of the river bed. The author distinguished whether the stretches were tidally influenced or not. In both cases the importance of the bedload transport for the processes of bed formation could now be explained in detail. Fargue considered it as important in this context to know as exactly as possible the sediment yield as well as its relationship to the cumulative run-off, a figure which he called torrential coefficient. The downstream-decreasing gradient is controlled by this "torrential coefficient" and the bedload attrition which gradually transforms larger particles into sand and silt. We recognize here certain ideas developed only shortly before by Sternberg (1875) as the object of his studies of the longitudinal and transversal profiles of rivers (cf. Chap. 4.2.4). From these considerations those deposits of sediment have to be excluded which have become stationary for whatever reason. The results of this study which, interestingly enough, avoids all mathematical discussions, can be compiled as follows:

The distance between the banks should not increase downstream at a uniform rate, but rather in keeping with the conditions of curvature. The width should increase downstream between two crossings at the same rate as the curvature until it reaches its maximum at the apex of the arc. Over the farther course the width should decrease in steps down to the next crossing. In the Garonne, as far as it is not influenced by the tides, the ratio between largest and smallest width in such a section is about $4:3$.

The displacement of pool and riffle on the crossing, as expressed by the first law, amount each to about twice the width of the river. The observation that the extremes of depth do not coincide with the apex of the arc or the crossing is ascribed by Fargue to a general law of nature. The action of forces onto inert masses generally engenders a temporal phase displacement just as the daily temperature curve lags behind the course of the sun.

The author concluded his study with the enumeration of certain unsolved problems and added with regret that river hydraulics was of all branches of constructional sciences the least advanced; and this although it is not less stimulating − not to speak of the scenic appeal of rivers − and the vast scope of the subject offers a wide field for personal imagination and initiative. Engineers should thus continue with the tedious task of observation and should condense the results obtained into rational syntheses.

Whereas these days the mathematical derivations of Fargue are virtually forgotten, the verbal descriptions of his laws have survived. The author had intended them mainly as a guideline for hydraulic engineers, but also for the student of river morphology to apply them with profit in his attempt to explain the processes of bed formation.

5.6   Regime Theory

Additionally to the well-established physical analysis of the flow processes and bedload transport, around the turn of the century empirical-statistical methods were first introduced being known in their entirety as the regime theory. Having its home in Great Britain and the Commonwealth, this is still acknowledged there as a well-proven tenet of hydraulic engineering. An overview of the basic concepts of the regime theory will be presented in the following on the basis of a compilation by Zeller (1965).

The term regime cannot be defined in a generally acceptable form. Terminologically interpreted, a river of the regime type has to be visualized as a stream flowing in its own alluvials.

A river or a river reach "in regime" indicates no long-term permanent modifications of its geometry, but is in a state of equilibrium or permanence. In other words, "to be in regime" means that the various parameters characterizing

Table 18. Exponents of regime equations

	Exponent	River cross-section	River reach
Water depth h	α	0.40	0.40
Channel width b	β	0.26	0.50
Flow velocity u	γ	0.34	0.10

a stream maintain well-defined relations to each other so that neither deposition, nor erosion, nor other variations of its bed will occur. The regime theory postulates registering measured values for parameters such as channel width, water depth, bed gradient, discharge, and flow velocity. Their mutual relationship is expressed by the so-called regime equations, the most important of which are:

water depth	$h = A_h Q^\alpha$	(100a)
channel width	$b = A_b Q^\beta$	(100b)
bed gradient	$J_s = A_j Q^\gamma$	(100c)
flow velocity	$v = A_v Q^\delta$	(100d)

According to the law of continuity there is

$$A_h A_b A_v = 1 \; ,$$

and

$$\alpha + \beta + \delta = 1 \; .$$

It must be taken into account whether the equations apply to a river reach and its immediate vicinity or to a prolonged stretch of a river, as in the latter case the increased size of the drainage basin has to be considered. The exponents of the regime equations cover a wide range, depending on the drainage basin and local climate. The approximate averages which may be anticipated are presented in Table 18.

Concerning the selection of a representative discharge, Q it will be of advantage to start from a idealized approach with respect to the frequency of recurrence. Whereas the older authors used the bank-full discharge, more recent theory adopts the discharge capable of transporting the largest sediment load above $d = 0.06$ mm, or even better, the effective discharge. This is the discharge which is able to transport the annual bedload yield within the period over which, during the average year, bedload transport takes place (cf. Chap. 4.2.7). The coefficients A must be determined separately for each drainage area, supported by measurements in test reaches.

By the inclusion of rivers in the regime theory, which originally had been conceived for canals, the importance of sediment transport became recognized. The methods used to outline its influence are, however, not on a par with the shear stress theory, and thus hardly satisfactory. Without going into details, it can be stated that the regime equations are applicable only to channels with fully developed bedload transport. Rivers with natural bed armoring or with a general lack of bedload motion, as well as mountain torrents, cannot be included here.

Newer researches attempt to reconcile these two theories and can furnish worthwhile results by this approach. Regime equations for h, b, and v were, for example, derived with the aid of the law of bedload transport of Meyer-Peter (cf. Chap. 4.2.7). The results were discussed in detail by Zeller.

Numerous regime equations have been developed as an aid to the planning engineer, offering average values for coefficients and exponents. In order to assure a wider range of applications, four different types of bed materials are distinguished.

Group	Properties of channel
A	coarse-grained, noncohesive material
B	sandy material
C	sandy bed, banks slightly to strongly cohesive
D	slightly cohesive materials

The following equations apply in the metric system:

1. Wetted perimeter

$$U = 4.75\,Q^{0.512} \quad \text{(Group C)} \tag{101}$$

2. Hydraulic radius

$$R = 0.273\,Q^{0.361} \quad \text{(Group A)} \tag{102a}$$

$$R = 0.476\,Q^{0.361} \quad \text{(Group C)} \tag{102b}$$

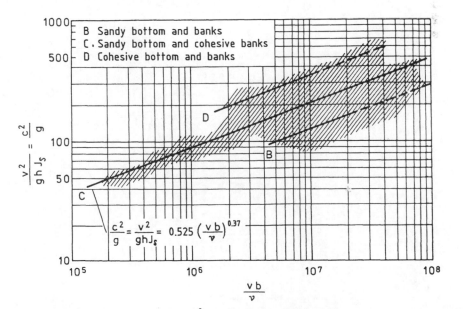

Fig. 5.24. Relationship between $v^2/g\,h\,J_s$ and vb/v. (Zeller 1965)

3. Flow velocity

$$v = 10.8 \, (R^2 J_s)^{0.286} \quad \text{(Group A)} \tag{103a}$$

$$v = 9.33 \, (R^2 J_s)^{1/3} \quad \text{(Group B)} \tag{103b}$$

$$v = 10.8 \, (R^2 J_s)^{1/3} \quad \text{(Group C)} \tag{103c}$$

4. Filled cross section

$$F = 0.94 \, Q^{0.873} \quad \text{(Group A)} \tag{104a}$$

$$F = 2.24 \, Q^{0.873} \quad \text{(Group C)} \tag{104b}$$

The relationship between $v^2/g h J_s$ and $v b/v$ can be obtained from Fig. 5.24, in which the range covered by the data of the various authors is indicated by hatching.

When using the various equations mentioned, the system is "overdefined". The partial results obtained must be checked for internal contradictions and weighted against each other.

When a rapid overview of the hydraulic and geometric problems of streams is required, the regime theory will furnish valuable aid. The simple make-up of the equations must, however, not lead to an excessively schematic application. A fruitful use of the formulas, on the other hand, presupposes a wealth of experience resulting from long years of observing rivers and their formation processes.

6 Classification of Rivers

6.1 Drainage Basins, Divides, and Their Structural Controls

The catchment area of a stream or river, i.e., the drainage basin − not necessarily the area of precipitation − is not a level plane. It is rather defined as "the area, measured in horizontal projection, from which the water flows to a certain location" (DIN 4049), i.e., as a portion of the earth's surface displaying a declivity to a certain point. The difference between the true surface area in the field and its projection over a map can be considerable and increases with growing angles of inclination. Keller (1962) presented the following example:

declivity of plane	0°	10°	20°	30°	40°	50°	60°
actual surface area in km^2	1.0	1.0154	1.0642	1.1547	1.3054	1.5557	2.00

At 60° inclination the true surface is thus twice the map projection area. These differences have to be taken into account accordingly during morphometric investigations (Range 1961). This is of greater importance for small drainage basins, especially in mountainous areas, than for larger areas. For river systems the differences frequently have to be disregarded as there is often a lack of reliable maps. The true surface, however, must not be neglected in measurements of evaporation and in investigations of the bedload transport, as erosion increases in steeper country. With some restrictions this naturally also applies to very large areas. It does not apply to determining the annual precipitation, as this relates to the horizontal plane.

The divides marking the boundaries between the drainage basins are usually also morphologically recognizable. Their exact delineation is nevertheless not always easy to define, as the superficial and subterraneous drainage areas in many cases do not coincide, the geomorphological (superficial) one being distinct from the hydrological (subterraneous) one. Their boundaries deviate from one another when the terrain is made up of jointed limestone prone to karstification, or of sandstone, as can be frequently observed in plateau country. The most prominent example is the capture of the source area of the Danube by a tributary of the Rhine (Chap. 7.1.2.2). It is almost impossible here to exactly define the subterranean drainage basin: for the Rhine it has become larger than the morphologically defined boundaries, and for the Danube smaller. The affected areas are, however, small in relation to the total drainage basin.

Exceptions from the morphologically clearly defined drainage basins are represented by valley divides or watersheds which can change with time. The water can run off to two sides, as can be observed locally in the glacial valleys of the

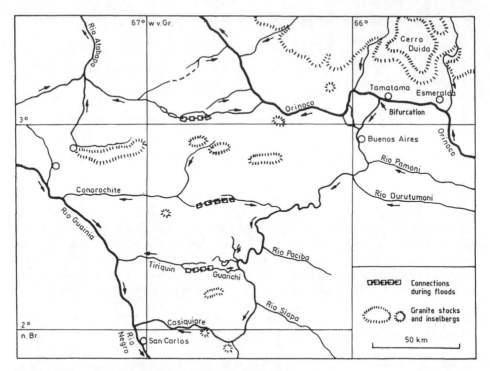

Fig. 6.1. Bifurcation of the Orinocco. (After Vareschi 1963)

North German plains which have now been abandoned by their original large rivers. Marcinek (1975) reported that the ancient melt water valley of the Red Luch east of Berlin may drain away to the Oder or to the Elbe river system. Bogs also drain at times into different areas, i.e., the position of the divides can change rapidly, depending on intensity and duration of the precipitation. Diffuse boundaries are frequently found in areas which had been subjected to glacial overprinting in relatively recent times. River bifurcations occurring on only slightly inclined plains can also be interpreted as valley divides. The most prominent example is the Casiquiare in Venezuela, through which the Orinocco also feeds the Amazon basin. During floods, additionally a number of streams in the vicinity of the Casiquiare also start draining into the direction of the Rio Negro-Amazon. A number of these streams do not completely succeed in this and only reach a neighboring tributary of the Casiquiare (Fig. 6.1). The intermixing of the turbid waters of the Orinocco with the clear black-waters of the Rio Negro also has biological consequences, as the two biotopes exhibit considerable differences (Vareschi 1963, cf. Chap. 4.3.3). In these cases outlining the divide is virtually impossible, it runs as it were along the bottom of the river bed. A less-known example is described by Louis (1961) from Central Africa, the bifurcation of the Logone. This river usually flows in the direction of the Chad basin, but during floods it may be driven off to the Benue and thus to the Congo basin (cf. Chap. 7.1.1).

Divides are not always clearly defined where two or three rivers have a common delta, and especially during floods the boundaries can become diffuse. Joint deltas are formed, for example, by Rhine and Meuse, Ganges and Brahmaputra (Bangladesh), Euphrates and Tigris (Shatt-el-Arab), Amazon and Rio Tocantins. The heavy monsoon floods in Bangladesh result in a huge area of inundation.

The discharge of the Amazon usually includes that of the Tocantins, just as the drainage basins of the two are considered together and the Tocantins is taken as a tributary of the Amazon. On the same premise, the Meuse should be considered as a tributary to the Rhine, a situation that actually was developed during the late Glacial and early Post-Glacial periods, before the sea reoccupied the area of the present North Sea. At several times in the past even the Hwang-ho and the Yangtze-Kiang had a joint delta, separation and junction having occurred repeatedly.

A drainage basin is not only outlined by the divides, but there are also local divides found between the various tributaries to the principal river; right up into the smallest creeks. Different from these are the continental divides which not only delimit entire drainage basins, but also direct the run-off of entire continents to certain oceans. This is referred to as peripheral run-off directed seaward. In several continents these divides at the same time form the backbone of large areas of inland drainage unconnected to any ocean, the so-called centripetal drainage. The areas not draining into the oceans account for about 30×10^6 km^2, or about 20% of the land surface of the earth. Here the genuinely nondrained dry regions must be distinguished from those with inland drainage, a difference that is frequently overlooked. The drainage-free nature of such regions can mostly be ascribed to climatic reasons and only rarely to the lithology of their bedrock. Save for large rivers passing through them, they contain only dry valleys, i.e., ephemeral streams, just as large portions of the Sahara are without any rivers or streams at all. The systems of inland drainage consequently should be specified as "without connection to the sea". Smaller areas completely without discharge outlets are found in peripheral river systems such as the Nile, Niger, Ob, etc. as well as in centripetal ones (Caspian Sea basin, Chad basin, Great Basin of the USA, etc.). References are found in, e.g., Maull (1958), Louis (1961), Keller (1962), Weber (1967), Wilhelmy (1972, 1974), Marcinek (1975).

The configuration of an area can show quite irregular forms, depending on whether drainage comes essentially from mountains, plateau country, plains, or glacially overprinted regions. The shape of the drainage area, as shown on the map, thus indirectly reflects the geological and tectonic lay-out. Details of this type, however, can rarely be derived from maps (cf. Sects. 6.2 and 6.3), but the morphometric parameters (Range 1961) and water control information can be obtained thereby (Brenken 1959; Klostermann 1970; Leder 1970). It also becomes obvious that the ideal plan displaying erosion, transition, and deposition areas is almost never encountered in outline and if so, only on a small scale. It is also rarely found in determinations of channel frequencies, i.e., the number of streams per unit area, and can thus only be considered as special type, not suited for the classification of rivers and their drainage basins.

In addition to the calculation of the topographic and morphometric parameters of a given area, it is also worth investigating whether in the lay-out of such an area not only climatic and geological forces are also "hiding", but also trends more or less independent of the former two which can be assessed topologically. Such studies are frequently carried out concurrently with determinations of channel frequencies:

Horton (1932) introduced a shape or contour factor, i.e., he determined the ratio of area of the drainage basin to its longest dimension as taken from the mouth to the farthest removed source.

The circularity ratio compares the area of a basin with that of a circle of the same size (Miller 1952 in Morisawa 1985).

The elongation ratio is based on the diameter of the corresponding circle and the longest dimension of the drainage basin, in parallel to the main discharge direction (Schumm 1956); Chorley et al. (1957) introduced the lemniscate ratio $K = L^2 \pi / 4A$, in which L is the largest diameter and A the area of the drainage basin.

Statistical grain size parameters such as skewness, kurtosis, and standard deviation (cf. Chap. 4.2.2) and grain shape analyses were used for comparative purposes, for example by Ongley (1970) and Jarvis (1976a). The methodical approach of recognizing distributions and irregular shapes is similar and employs the Fourier analysis for this purpose. The latter may also be used for the determination of the configuration of the perimeter, the circumference of the drainage basin.

McArthur and Ehrlich (1977) have introduced a formula for the calculation of the average radius of an area with which they were able to delineate characteristic differences in the shape of areas. However, these and the other mathematical-topological methods will not allow the interpretation of purely geometric factors, because controlling mechanisms such as the configuration of an area can also be explained by lithological factors. This applies just as well to all other highly theoretical methods of determination of channel frequencies.

Morisawa (1958) already observed in comparisons that the elongation ratio correlates best with the hydrological situation, and that with the circularity ratio differences in shapes can be characterized, whereas the lemniscate ratio is less suitable.

In addition to the topographic-cartographic presentation which is used for most investigations, individual diagrams (Fig. 6.2) can be used for a rapid assessment giving schematic information on length and size of an area and on its internal and external configuration.

6.2   Drainage Density, Channel Frequency

The distribution of water courses within a drainage basin, whether they be perennial or ephemeral, is not accidental, either with regard to the number or to the location in the countryside.

The aim of the investigation of channel frequencies is to obtain a relatively rapid indication of the geomorphological character of an area. It serves especially in poorly accessible country as a first and frequently rather good information on

Configuration of drainage
basin of the Inn

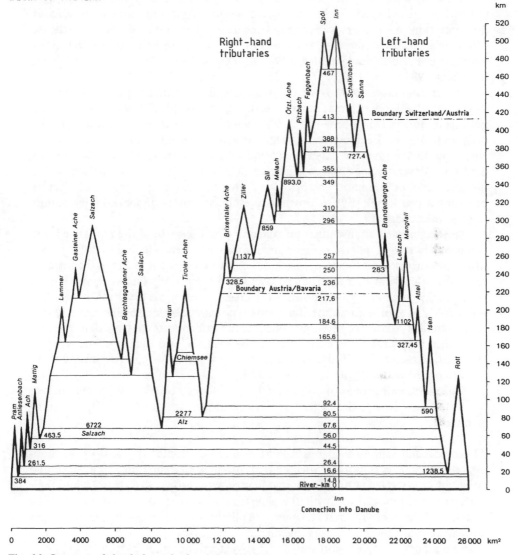

Fig. 6.2. Structure of the drainage basin of the river Inn

type and distribution of the rocks encountered as well as of their erodibility and
tectonic structure (Kronberg 1984).

The accuracy of the assessment depends on the quality of the maps available
(cf. Chap. 5.2.1) or on the scale and the quality of any aerial photography. A deci-
sion has to be made furthermore whether an investigation is restricted to the
perennial water courses or whether the additionally present dry valleys should
also be considered. It is possible to determine two values: the present channel frequency or the general valley frequency (Horton 1945; Dingman 1978).

Many valleys were formed in the geologically recent past when a different hydrographic situation prevailed. Under certain conditions they can become partly or wholly reactivated, leading to revived erosion. As a consequence, at present the total number of courses is mainly considered. The transition from a mature valley to a shallow depression is a flowing one, but in most cases a satisfactory distinction is possible (Strahler 1954; Morisawa 1958; Leopold et al. 1964; Kronberg 1984).

The determination of the channel frequency allows a preliminary statement as to the erodibility of an area, but the recording of the total number of channels permits a more uniform picture across climatic zones. Furthermore, there will be no need to answer the question frequently put in earlier times as to whether larger rivers should be given a larger weighting. The total frequency of the courses developed depends, as in the case of rivers (cf. Chap. 5.1), on the nature of the bedrock and on climatic factors, with one or the other group of factors predominating. It will thus be possible to establish points of special interest during the investigation.

The accuracy of an investigation depending on the scale of the maps used, the frequency is always determined for small drainage areas which are easily recognized and surveyed on maps or aerial photos. A regional comparison then results from the evaluation of numerous subareas and their statistical analysis.

Here this method finds its limits, as the results of one area cannot be transferred directly to another, and the same rock type may react quite differently in different climatic zones. However, morphoclimatic information may be derived from a uniform comparative determination of frequencies in different areas, whereas the consideration of only the permanent water courses would give an incorrect picture of the bedrocks and their erodibility.

Various possibilities of interpretation will be briefly discussed below, separated according to the factors bedrock (geology, structure) and climatic morphology.

Geological and Structural Aspects. For the interpretation of the network of channels recorded, it is important to note that and how the partition, frequency, and orientation of the trend lines of discharge vary with the types of bedrock and their different erodibilities, and with the geological structure of the subsurface. The analysis of the discharge system of an area allows conclusions as to the rock types present and to their spatial arrangement.

In regions of deep erosion and/or dense vegetation, and in particular in poorly accessible tropical rain forests, the mapping of this network is the foremost method for recognizing lithological and tectonic structures.

The analysis of the drainage network according to frequency and especially to the orientation of the streams (cf. Sect. 6.3) supplies information also in areas where outcrops are covered by weathering profiles or by younger deposits of varying thickness. In debris and alluvial material of coastal plains and inland basins frequently the trends of discharge are tracing zones of jointing or faulting in the bedrock. Performance of the analyses on maps and aerial photos can, however, never quite replace verification by field work.

Literature for the geological and structural aspects: Parvis (1950); Chorley et al. (1957); Murawski (1964); List (1969); List and Helmcke (1970); Gustafson (1973); Drexler (1979); Kronberg (1984).

Climatological Aspects. In morphoclimatic investigations the recording of all erosional channels is of special importance. The immediate channel frequency in semi-arid and arid regions is extraordinarily small. Nevertheless, and especially here during the infrequent but locally excessive downpours, heavy erosion will take place. However, when the total frequency of channels, precipitation, and vegetation are related to each other, some clear correlations can be found.

In principle, channel frequency and precipitation rate may be related to one another by the mean annual precipitation and the intensity of precipitation, the latter being expressed for a period of 24 h. Also in these cases the various authors mentioned below have selected numerous, usually small subareas and determined the total number of streams per unit area from the maps.

Such investigations showed, for example: As expected, the frequency diminishes with increasing precipitation as the plant cover in areas of stronger precipitation tends to be denser and thus will inhibit dissection of the landscape by valleys. Abrahams (1972) found the density maximum at above 1000 mm mean annual precipitation. In the rain forests of Guyana, Daniel (1981) observed the maximum at a precipitation of 3000 mm a^{-1}. As a logical consequence the frequency increases with decreasing precipitation as the protection provided by the vegetation becomes smaller. Abrahams gave a frequency maximum at only 280 mm mean annual precipitation.

The intensity of precipitation is considered important as in areas of low vegetation cover it can lead to stronger corrasion and dissection due to the greater erosive force (Chorley and Morgan 1962). Gregory (1976) found a direct relationship between the frequency and a precipitation intensity index, whereas Abrahams and Ponczynski (1984, 1985) observed only an indirect relationship between frequency and intensity, but a direct one between the quantity of precipitation and its intensity. Areas with comparable mean precipitation do not necessarily have the same channel frequency. Channel frequency and sediment load of the flowing channels are not necessarily interrelated either.

In the methods selected for the widely separated morphoclimatic investigations of channel frequencies, the influence of geological and structural factors is frequently not discussed. Mostly only the vegetational cover varying with the climate is taken into account. The geological parameters, however, also have to be considered. The three main factors, lithology, climate, and vegetation, here exert the same influence as is valid for the general geometry of streams (cf. Fig. 5.1).

Abrahams (1972, 1984); Abrahams and Ponczynski (1984, 1985); Carlston (1966); Chorley and Morgan (1962); Cotton (1964); Gregory (1976); Gregory and Gardiner (1975); Madduma Bandara (1974); Melton (1958); Morgan (1973); Daniel (1981); etc.

Channel frequencies were used early on for morphological investigations. Gravelius (1914) discussed a number of pertinent papers. In Europe and the USA diverse methods were developed concurrently which, although interesting in their approach, are mostly only of historical value. What has remained generally is the determination of the total length in the unit area, now almost exclusively given in $km\,km^{-2}$. The more important European and North American approaches will be discussed below.

European Studies. One of the first papers is by Belgrand (1873, cited by Gravelius 1914). He attempted to explain the differences between certain landscapes in the Seine basin with the aid of i.a. channel frequencies. In the German-language literature Penck (1894) first pointed out the importance of this problem. He also presented numerical values as a measure of the frequency by stating that in the central Alps the distances between larger rivers are 5–6 km. These are joined by tributaries at 2–3 km intervals, which in turn are joined by smaller streams spaced at about every 250 m.

Within a comparatively short period of thime there were papers published by Neumann (1900), Spöttle (1901), Feldner (1903), Suerken (1909), Puls (1910), Wolff (1912), Rasehorn (1913), Schäfer (1913), Sauerbier (1918), Böhmer (1922), Fluck (1925), Trübswetter (1930), Gönnenwein (1931), Gutersohn (1932), etc.

These authors were essentially applying three different criteria for the determination of the channel frequencies of a certain area:

1. Neumann was the first to systematically determine the channel frequency as the "ratio of the aggregate length of all natural water courses of the respective drainage basin to its surface area", or, in short, as the length of the water courses per unit area, i.e., per km^2. This method can be considered as the most practicable one. The paper by Spöttle also deserves mention in this context. He had been assigned the task of determining the total length of the flowing water courses of Bavaria as rapidly as possible. For this purpose he recorded the channel frequency according to Neumann for selected areas underlain by a certain geological formation. By suitable extrapolation he applied the resulting figures to the individual geological regions and then to the whole of Bavaria.

2. Suerken, in his investigation of the Münster/Paderborn basin of northern Germany, where subterraneous and surface divides are difficult to locate exactly, used instead of Neumann's method a trapezoid as the unit area, the base lines of which extend over 2' of latitude whereas the sides measured 1' of longitude each. This resulted in an average unit area of about 4.2 km^2. The drawback of this method is that the area represents only an average figure. This yields only small differences if the areas extend over a small range of latitudes only, but could lead to considerable differences for areas of large north-south extent.

    Trübswetter (1930) used the trapezoid proposed by Claas (1922, in Trübswetter) which extends on all sides to 1' longitude and also latitude. The results obtained by him were more accurate than those of Suerken.

3. Feldner used the areas separating the various water courses. This method results in "peninsular"-shaped areas open to the divides. The locations of the springs he connected by a straight line. According to Feldner, the channel frequency of an area is expressed by the different sizes of the individual areas ("river network mesh"). The larger these areas are, the smaller the channel frequency will be. The method was modified by Rasehorn, who mapped the

divides in greater detail and who then did not determine the size of the individual areas, but rather related their number of the size of the drainage area. He quantified the channel frequency by the average size of the meshes, the frequency being greater for smaller average areas. Rasehorn furthermore followed suggestions by Penck and Gravelius and determined the mean distance between rivers which led to essentially similar results. The Feldner-Rasehorn method is quite independent of the determination of the stream lengths. The differences in the channel frequency are often more obvious in the results obtained with this method than with those of their subdivision into grid squares. Wolff, Böhmer, and Puls applied similar methods, the latter using the standard survey map size as the unit area (1:25000).

However, the subdivision into squares was generally preferred, although not always with a side length of one kilometer. Puls furthermore applied a method that had initially been developed for the determination of the network density of a railway system. In this the total area investigated was converted into grid squares. Fluck as well as Schaefer preferred a length of 2 km. The final figures for the channel frequencies are, however, always given in the usual dimensions of km km^{-2}. Gönnenwein also determined the frequency for a unit area of 1 km^2, but in the graphical presentation then used grid squares of 3 km side length for reasons of clarity (Fig. 6.3).

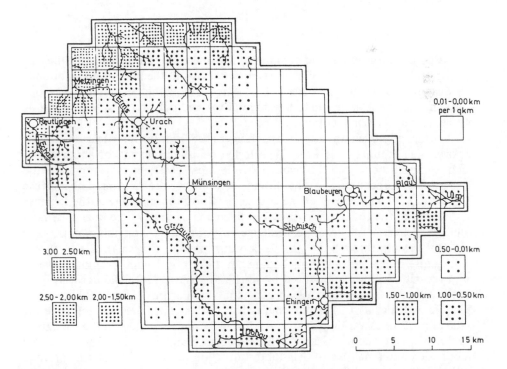

Fig. 6.3. River density map of middle Suebian Alb. (After Gönnenwein 1931)

River density [km/km²]

☐ 0 - 1   ▓ 15 - 20
⊡ 1 - 5   ▤ 20 - 25
▨ 5 -10   ▦ Above 25
⊞ 10 - 15 ↴ Sub-areas

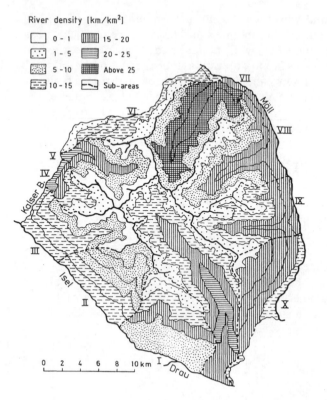

Fig. 6.4. River densities. (After Paschinger 1957)

Based on the method of determination of mean coastal distances, Fluck as well as Trübswetter used the mean stream distance, i.e., the lines of mean boundary distances to the nearest water course. The cartographic presentation in both cases showed the relationship to lithology of the bedrock and to structure just as clearly as did the channel frequencies.

Another approach was used by Paschinger (1957) for his investigations directed over high mountainous areas. He selected elevation intervals of about 400 m each, to which he related the channel frequencies. These intervals represent the approximate vertical extent of the climatic zones in the central Alps. The increase in area and the lengthening of the course resulting from this due to the increase in declivity of the country have to be considered accordingly. The final statement will be the channel frequency per drainage area of a larger mountain water course which is passing through the various contour intervals (Fig. 6.4).

A more recent method using squares of 1 km grid length, extending over a larger framework, was introduced by Karl and Höltl (1974). The Bavarian part of the Alps was to be given a landscape analysis. For this purpose the area was overlaid with a grid of 1 km unit length. This grid included 5200 squares covering 4800 km² of the Bavarian part of the Alps and some hilly areas in the forelands. For each square about 20 ecologically important parameters were recorded, and subdivided into a number of long- and short-time variables. Long-term variables included the channel frequency. With this the channel frequency as an eminent

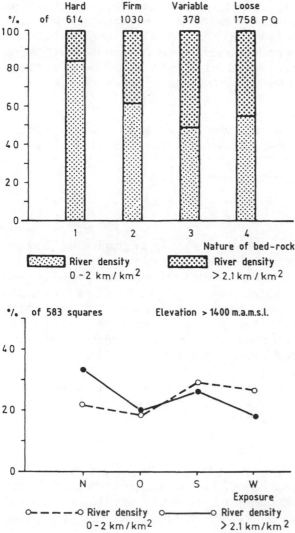

Fig. 6.5. River densities. (After Karl and Höltl 1974)

grid feature was accorded its proper place in the assessment of the ecological character of each area. In the analysis the parameters were compared with each other by single and multiple regression correlations. In this way it was possible to illustrate the stress potential and stability of a highly differentiated region (Fig. 6.5).

North American Studies. Literature: Gilbert (1880 in Schumm 1972); Davis (1890); Horton (1932, 1945); Strahler (1957); Leopold et al. (1964); Morisawa (1957, 1958, 1962, 1963, 1985); Shreve (1966); Woldenberg (1966); Milton (1967); Ranalli and Scheidegger (1968); James and Krumbein (1969); Mock (1970); Jarvis (1972, 1976c); Werner and Smart (1973); Mark (1974); Onesti and Miller (1974); Warntz (1975); Smart (1968a, b, 1969, 1978); Flint (1980); Wiltshire and Hewson (1983), etc.

In the USA channel frequency studies have been frequently pursued. Concepts were based on the work of Gilbert (1880 in Schumm 1972) and on the system of Davis (1890). The proposals of Horton (1932, 1945) which are partly based on Gravelius (1914) have gained considerable acceptance. Horton, too, was looking for relations between erosion and channel frequency as well as between discharge and channel frequency, and in this he started from the simple relation

$$D = \Sigma L / A \ .$$

In this equation, D represents the cumulative length L of all channels divided by the drainage area A, initially presented in miles per square mile but now almost exclusively by km km^{-2}.

Schumm (1956) preferred the reciprocal value by setting

$$C = \frac{A}{\Sigma L} = \frac{1}{D} \ .$$

The value C he called constant of channel maintenance, originally given in square feet per foot, now mostly by m^2 m^{-1}, considering it as a measure for the erodibility of an area. Regions with erosion-resistant hard rocks or with permeable rocks, also with dense vegetation cover, have high C-values and low densities. On the other hand, a low C-value results from high densities over soft rocks with low permeability and little vegetation cover and correspondingly high rate of surface run-off. The factor may be applied, but in its interpretation also a detailed description of the landscape is required, together with a knowledge of the morphoclimatic ambience.

Carlston (1963) used hydrogeological parameters for preference. The density D is related to the transmissivity T of soils and rocks. Based on a groundwater model, Carlston established the following equation

$$D^{-2} = \frac{8 h_0 T}{W} \ .$$

In this, h_0 is the vertical distance between the groundwater level and the bottom of a stream at the divide and W the rate of recharge. D^2 correlates inversely with the base flow Q_B and is further proportional to the factor Q_P, i.e., the discharge contribution per unit area.

Chorley et al. (1957) proposed an approach by climate and density of vegetation. He introduced the parameter I which represents the amount of vegetation divided by the product of rainfall amount and rainfall intensity. The channel frequency is then the reciprocal of this value, or

$$D \sim \frac{1}{J} \ .$$

Channel frequency and structure of an area were plotted against each other by Strahler (1954). The texture ratio, $T = N/P$, is plotted on the abscissa, and on the ordinate the channel frequency, $D = \Sigma L / A$. P represents the perimeter of the drainage area whereas N "is the number of notching crenulations on the contour having the maximum number". In a logarithmic presentation Strahler found a

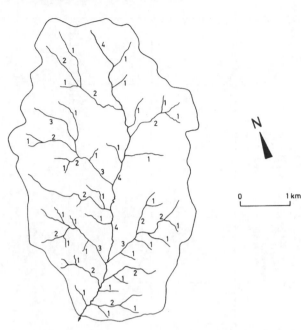

Fig. 6.6. Stream ordering after Horton. (In: Leopold et al. 1964)

near-linear correlation, starting from low density and low or "coarse" T-values to high densities with high or "ultra-fine" T-values. The results are comparable to those obtained with the aid of morphoclimatic factors.

Horton (1945) attempted better efficiency by emphasizing the topological features of a drainage area. For a start, he gave a grading or order number to every channel in the hierarchical structure of a drainage basin. With this he was able to establish relationships between river order and lengths of courses, between river order and size of the respective drainage basins, as well as with number of streams of a certain order number. The resulting diagrams have become established as Horton diagrams in literature. Such analyses can be carried out for very small areas or whole countries, depending on which stream size is used for starting the count. The order of counting starts with 1 for the smallest stream observed and ends at 11 or 12 (Fig. 6.6). In each case the map scale is the determining factor. The first order on the respective map is accorded to streams without tributaries, the second order to those which have first-order tributaries. Most appropriately, the second order is traced up to the source of the strongest first-order stream, but some latitude is given. Streams of the third order have only first or second order tributaries, etc.

Strahler (1957) modified this system by starting the next-highest order at the confluence of two tributaries of lower order, in this way obviating the need of tracing one of these back to its source.

Leopold et al. (1964), on the other hand, supported the original method. In this case it is necessary to decide upon whether areas dewatering directly into a higher-order stream without themselves having first-order creeks are credited to the first order to whether they are included only in the compilation of the whole drainage area.

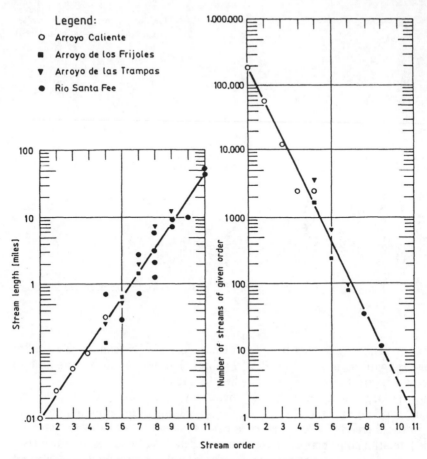

Fig. 6.7. Horton's analysis of various rivers. (Leopold et al. 1964)

The Horton analysis in Fig. 6.7 illustrates the geometric relations between the river order (always on the abscissa) and the corresponding number of channels, their aggregate length, or the size of the areas on the logarithmic ordinate. As an additional figure, Horton introduced the bifurcation ratio R_b. It is the ratio of the number of channels of a certain order to the number of channels in the next-highest order or to the average of a certain area. An example is presented below:

Order	Number of channels	R_b
1	32	—
2	10	3.2
3	3	3.3
4	1	3.0
		3.2

The ratio may also be derived from the slope of the regression line "number vs. order" in Fig. 6.7. The value for R_b tends to lie between 2 and 4, whereas for the arid southwest of the USA figures of about 3.5 have been reported.

When investigating very large areas, the scale of the maps used can result in considerable changes. Leopold et al. reported that at a scale of 1:24000 an eleventh-order river, i.e., a very large one with a relatively dry drainage area, will have about 200000 first-order streams. At such orders of magnitude the streams are non longer counted individually, but rather obtained from the right-hand diagram of Fig. 6.7 by extending the regression line to the largest order (11) on the abscissa. In a similar manner the summated length of streams of a certain order can be obtained graphically from the diagrams of Fig. 6.7.

On the scale mentioned the Mississippi at its mouth would have the order number 12, whereas on a scale of 1:62500 it would be 10. If this order 10 is used for the largest river of a country, as proposed by Strahler, all its other rivers will be of a lower order. Using this approach, Strahler's method results in a cumulative length of about 3.2 million miles (about 4.96 million km) for the rivers and streams of the USA.

This system of geometric-topologic ordering has been criticized repeatedly. It does not, amongst other points, allow for the possibility of changes exerted by a tributary coming in, but only continues with the order numbers. In total, the system is rather schematic, leading to a considerable loss of information.

Shreve (1966) and other commented that Horton's parameters are only related to each other, that they give too much room to statistics, and that they do not reflect the real characteristics of the area. The relations observed between the parameters of discharge are, for example, directly affected by the drainage area, implying that when the area increases and with it the discharge itself, the number of streams will increase together with the order number. It can thus be stated that the area is proportionate to the order number.

There have been a number of attempts to improve the Horton-Strahler system or to develop alternatives. A few of these new developments will be referred to below. For a more detailed compilation reference is made to Morisawa (1985).

Woldenberg investigated the possibility of considering a river system from the point of view of its energy content. With the aid of the bifurcation ratio and the central place theory, he assumed that even under the influence of restrictive geological factors the drainage areas will act and develop as a hexagonal lattice system, even when they are not really hexagonal. The hexagonal arrangement of bodies such as the densest packing of spheres and hexagonal lattice systems are generally known in science. According to Woldenberg, river systems are established in such a way that always the shortest distances are achieved and with this an equilibrium between the lowest amount of work and the entropy of the system. The energy losses of the river are thus minimized by the hexagonal configuration of the network.

This theory cannot be rejected off-hand, as a river in many cases will actually consume the energy minimum (cf. Chap. 5). However, Woldenberg restricted his investigations purely to counting on the map sheet, and in this way he introduced additional uncertainties.

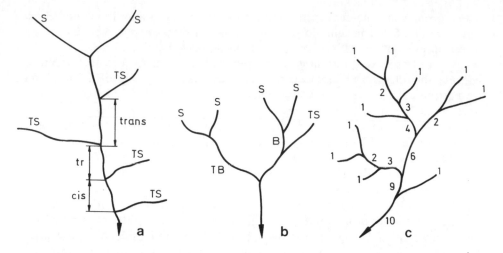

Fig. 6.8a–c. Basic drainage patterns (from various sources)

Various authors used different parameters for counting. Shreve (1966) introduced the term "link" for the stretches of a stream between the junctions of successive tributaries. He distinguished two types: (1) exterior links (source creeks), which extend from the source to the first tributary, and (2) interior links, which represent the stretches between two tributaries or the last tributary and the mouth of the stream considered.

A network with n sources has $n-1$ nodes and $2n-1$ links of which n are exterior and $n-1$ are interior. Links are designated as of a given magnitude which increases with each tributary (Fig. 6.8). The magnitude of a link is equal to the number of sources above it.

The disadvantage of this method lies in its restricted application to rather small areas as very high order numbers would be rapidly reached. Smart (1978) proposed a binary method of counting in order to take this problem into account. The authors suspect a random configuration of the network when there are no geological factors of control and consequently a stronger control is introduced by the precipitation rate.

James and Krumbein (1969) additionally recommended the use of the terms cis-links and trans-links. Cis-links are stretches of a stream between two successive tributaries coming in from the same side, whereas trans-links are those between two successive tributaries from opposing sides (Fig. 6.8).

In hard flat-lying rocks, these authors observed more trans- than cis-links, the streams thus being joined by tributaries mostly from either side. In their sample areas they moreover found fewer short cis-links than short trans-links. They explained these observations by the tendency of the system to develop from a more or less random initial network to a finally more ordered configuration. The area thus "matures" with time and distance from the source to the mouth (Flint 1980). This would imply that geological factors such as a network of joints increase with time in importance with the size of the respective factor and of the area. A completely random development without the influence of geological or climatic factors is not possible, even it were initially random to some degree.

Mock (1971) proposed a similar system (Fig. 6.8), the counting method of which is somewhat related to that of James and Krumbein. Mock starts with the exterior links of the former as source links (S) of the order number 1. He further used tributary source links (TS) for streams of order numbers possessing a tributary with an order number below 1. The interior links of James and Krumbein are called bifurcating links (B) or tributary bifurcating links (TB). Areas with stronger gradient relief contain a larger proportion of TS-links, i.e., their proportion varies directly with the intensity of the relief. In this case also the development is controlled by bedrock lithology and climate.

Comparable results were obtained by Jarvis (1972, 1976) and Werner and Smart (1973). The "concept of topologic (link) distance" of the former is similar to the "path length" of Werner and Smart. The latter describes the number of tributaries from the mouth up to the selected junction farthest upstream, as measured along the most direct flow route.

The stretches can be exterior or interior, depending on up to which final link they are measured. The longest possible path in a drainage basin is the one from the mouth to the uppermost exterior link. This is also referred to as the network diameter. The larger the network diameter, the longer the links will be, and vice versa. Above a network diameter of $10-15$ the length of the links appears to remain more or less unchanged (Jarvis).

Jarvis and Werrity (1975) introduced the rather debatable parameter "source height P". It is defined as the ratio between the total number of source links p, and the number of sources N_1 or $P = \Sigma p_1/N_1$. A low mean source height and a low network diameter indicate a more compact source area, whereas high P and a larger diameter indicate an elongate shape of the drainage basin.

With the statistical evaluation of the measurements of stream directions in the drainage basin of a river which are almost exclusively controlled by jointing, strike, and dip of the bedrocks, systems of jointing can also be traced which would show only faintly through soil cover or overlying glacial sediments (Jarvis 1976; Flarity 1978 in Morisawa 1985; Cox and Harrison 1979).

The angle between a tributary and its main stream is also measured, a method that goes back to Horton as well. As shown by field measurements and experiments, such angles depend on gradient and discharge as well as on the bedload yield of the tributaries and eventually on the relief (Howard 1971; Mosley 1976), i.e., they have to be considered as a three-dimensional phenomenon.

The angles are subjected to changes with time and the development of the river system, the direction of change, however, not being predictable. Angles also appear to change with increasing order number in the Horton-Strahler system (Lubowe 1964; Flarity 1978 in Morisawa 1985) and become larger as the order number of the main stream increases.

A number of recent papers deal with the preparation of digital topographical models, or else digital elevation models (Band 1986). Their aim is to split up the area into basic units which are then subdivided according to exterior and interior links into exterior and interior basin areas, making up the total drainage basin.

At a later stage complementary recordings of the vegetation cover and of the land surface as a whole will be carried out in order to achieve a comprehensive quantitative geomorphology which would then allow better predictions about discharge and general developments in defined areas.

6.3 Drainage Systems, Types of Rivers

The tree-like branched river system is arranged hierarchically. There are, however, numerous more or less characteristic deviations which are caused by morphologic, structural, or climatic factors. Some types preferentially occur in certain climatic zones.

There are essentially four different methods to typify rivers:

1. Types of drainage systems. The starting point is the drainage system itself, the arrangement of the individual channels.
2. Topographic-morphologic types according to the characteristics of the catchment area (mountains, hills, plains).
3. Climatic types.
4. Hydrologic types (regime types).

6.3.1 Drainage Patterns

The presentation according to the shape of the plan view considers mainly the influences exerted by the bedrock formations. In this respect there are close relations to the determination of the channel frequency. The in parts rather conspicuous outlines of the drainage patterns were brought into a system by Gilbert (1880 in Schumm 1972), Schumm (1972), Marcinek (1975), Kronberg (1984), etc. The descriptions below follow essentially the presentation of the latter author, as these are based on lithology and structure, and thus universally applicable. They also initially disregard local peculiarities.

Certain basic types — according to subdivisions, density, and orientation — are repeatedly encountered. They are in clear association with certain rock types and structures or else their development and properties. Such factors may be: mineralogical composition and mineral size of the rock, their texture, permeability, solubility, the presence or absence of bedding, the bed thickness, outcrop width, conditions of layering, presence and orientation of joints, as well as type, strike, and dip of faults.

Rock type and structure essentially control what portion of the precipitation will run off on the surface and act erosively, as well as in which direction and in which courses. In this way the surface run-off will shape the lithology- and structural-related, both small- and large-scale morphology. This again underlines the importance of recording all channels in the determination of channel frequencies.

One can distinguish between drainage patterns which develop solely as a result of the rock type present and those which are additionally or mainly determined by tectonic structures.

Dendritic Type. The most frequently developed pattern, the tree-like branched type, belongs to the river systems essentially controlled by the bedrock. The main river corresponds to the trunk of a tree, the tributaries of various orders to its branches and twigs. This basic type thus is referred to as dendritic. In a dendritic river system the main course and the tributaries are established without much control, but still subject to the laws of channel geometry. The tributaries join the subsequent streams at highly diverse angles.

The dendritic type will develop in many varieties over most variable rock types such as clays, marls, shales, and sandstones, sometimes also on igneous rocks. Dendritic patterns differ from each other with regard to channel frequencies and subdivision of the network. There are wide- and close-dendritic patterns with a variety of transitions (Fig. 6.9a, b). Wide-dendritic patterns occur i.a. over flat-lying, coarse-grained, unjointed sandstones etc. with good permeability, and locally even over igneous areas. Over fine-grained, relatively impermeable and easily erodible rocks such as clays, fine-grained sands, marls, tuffs, loess, etc. close-dendritic patterns are developed. In such a delicate network the channel frequency is correspondingly higher. It is of little importance whether these rocks are flat-lying or folded, as long as their thickness is sufficient to allow the establishment of a separate drainage network.

Trellis or Angular Type. In this type the land surface is governed by the succession of escarpments and intervening valleys parallel to the bedding, resulting from the differences in weathering and erodibility of these alternations of lithology. This also includes the system of Davis (Sect. 6.3.2.2). Consequent streams and rivers flow across the bedding planes in the direction of their dip. The subsequent rivers then follow the direction of the strike, etc. The main streams and their tributaries exhibit more or less rectangular bends and arrangements.

Trellis-type river systems are developed not only on inclined tablelands but also on horizontally orientated more or less homogeneous massive rocks which are cut by parallel zones of faulting and by steeply inclined systems of jointing and/or subsidiary faults. The larger rivers follow the easily eroded major fault zones, whereas the tributaries are controlled by the joints or the subsidiary faults (Fig. 6.9c, d).

We here have the phenomenon that similar drainage patterns can be found over lithologically or structurally diverse bedrocks.

In any case, linear stretches and abrupt changes of direction of main streams and tributaries are typical for all areas in which the drainage pattern and specifically its orientation are governed by zones of jointing and/or faulting systems. Joints and faults can express themselves in parallel, rectangular, or sub-parallel orientation of the channels.

Radial and Annular Patterns. These are usually special, but highly characteristic types. Radial or annular drainage patterns are developed, for example, on volcanoes where the precipitation runs down the flanks in all directions. Radial patterns are also found over the centers of ascent of salt or granite domes. In such cases already small differences in elevation between the points of ascent and the

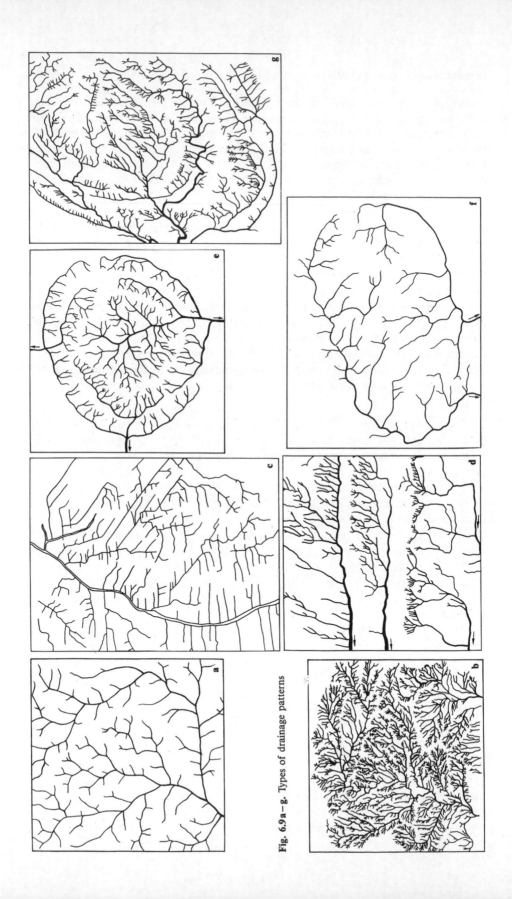

Fig. 6.9a – g. Types of drainage patterns

surrounding area suffice to establish radial (centrifugal) and annular drainage patterns. They are especially conspicuous within a larger system (Fig. 6.9 e, f).

In the areas of ascent, erosion selectively carves out the harder rocks. Over the rock sequences dipping away from the center a circular topography is developed with concentric valleys and ridges. The consequent streams here flow away radially over the bedding surfaces. They end in the larger subsequent streams which follow the circular strike of the ring structures until they reach a cross-cutting gorge through which they flow out and away from the ring structure. Characteristic combinations of radial and concentric lines may be recognized (Fig. 6.9 e).

Another special type is observed when the discharge runs centripetally from all sides into a former dome, since eroded out, and now having only one single outlet. However, this is found also in all other depressions on the surface of the earth, which can be filled, for example, by a lake with one outlet only or by a terminal lake. This represents the "centripetal subtype" of Marcinek (1975).

Mainly Structurally Controlled Types. Drainage patterns of the radial or similar types can also be found around plunging axes of anticlines and synclines. Figure 6.9 g illustrates a large anticline plunging to the left. The structure consists of an alternation of sediments with different erodibility. The annular streams are subsequent and they are jointed by the small consequent streams which flow on the inclined bedding surfaces or by the small obsequent streams in the direction opposite to them (cf. Sect. 6.3.2.2).

These attempts at a pattern typification are, strictly taken, only applicable to relatively small areas, as is the case for the channel frequencies. They must not be directly transferred to other regions without considering climatic conditions and therewith the vegetation cover, as well as elevation and exposure. Larger rivers are usually more of the wide-dendritic type.

A similar approach, of using the types of river courses as a criterion of subdivision, was applied by Fluck (1925) for the rivers of the Swiss Jura Mountains. The results are similar to the classification of Davis, an observation that is not surprising in view of the geological situation of these mountains.

6.3.2 Topographic-Morphologic Types

The topographic-morphologic types are subdivided according to their occurrence in the different regions. In this way purely climatic criteria are pushed somewhat into the background, although they always exert some influence on each type.

6.3.2.1 Torrents, Mountain Creeks, and Mountain Rivers

The explosive touristic development that has affected the Alps during the last decades has of necessity directed attention to damage by erosion through running water and avalanches. These have been caused or increased not least by human activities. Increased demands for safeguarding tourists as well as public and private property have resulted in preventive safety measures against torrents and avalanches and increased regulation of rivers. The basic research concurrently carried out has yielded numerous new data and also confirmed old observations and experience. The origin of torrents has received increased attention. We now dispose of much know-how on the interaction of natural causes and mankind on running waters in mountainous regions. This new insight is not always of a purely morphological character, but attributes the main importance to morphology, rock type, and vegetation. These investigations have created a separate terminology descriptive of the various types of torrents and mountain creeks. Due to differences in purpose, the official and the scientific terminology are not always in agreement.

Two approaches can be distinguished in these classifications:

The first attempts to record as many individual factors as possible in order to arrive at a concise picture of what is referred to as the "torrentiality", a not yet familiar term derived from the French-speaking parts of the Alps (Margaropoulos 1960; Messines 1964; Munteanu 1970; etc.). This classification, however, has not yet been accepted by forestry administration or torrent-regulating agencies.

The second approach gives prominence to the geological factors and classifies the streams according to the degree to which they are affected by human activities. The older papers, e.g., Stiny (1910, 1931) already provided a basis; younger publications are by Karl (1970), Stern (1971), Moser (1971, 1973), Kronfellner-Kraus (1973), Karl and Mangelsdorf (1971, 1975), etc. On their results are based the modern methods of torrent control which saw themselves confronted in practical work with the necessity of solving many questions, especially of a geological and morphological nature. This has resulted in detailed maps and several attempts at classification, of which the one by Karl and Mangelsdorf will be introduced as an example.

One such categorizing principle concentrates on the question as to practical man-induced effects, the other emphasizes the type of source of the bedload, the actual suppliers according to the definition of torrents by Karl (1970). The aim of such classification is to find out whether integral control measures would pay off in a certain case or whether any protective measures mooted are directed only at certain individual targets, unaccompanied by measures modifying the drainage area upstream.

This definition differentiates between torrents and mountain creeks. Defining all mountain creeks as "bedload-bearing" (Hampel 1969) appears to be an oversimplification. The definition by Messines (in Margarapoulos 1960) as "a small mountain water course flowing intermittently cr perennially steeply downhill, with sudden and tremendous rises of water level, the discharge and solid load transport of which shows large fluctuations", is not precise enough. "Small watercourse" is vague, and fluctuations in discharge and transport of solids are encountered in virtually all water courses in mountainous areas.

Karl defined: "A torrent is a water course with at least locally strong gradient, rapidly and highly variable discharge, and at times high transport of solids especially of bedload, these solids being derived directly from localized point sources (overhanging cliffs)" (cf. Chap. 4.2.3). This distinction from mountain

Table 19. Types of torrents in the Alps and similar young mountain ranges

Potential of influencing sources of solids	Torrents with expansive sources of solids in residual alluvial bodies. Discharge and erosion of solids influenced by vegetation	Torrents with stationary sources of solids – discharge and erosion of solids not influenced by vegetation	Torrents with stationary as well as expansive sources of solids – discharge and erosion of solids partly influenced by vegetation	Torrents only at times with own sources of solids – discharge and erosion not influenced by vegetation
Sources of solids in:				
Resistant igneous rocks		Torrents in resistant igneous rocks		
Resistent sedimentary rocks		Torrents in resistant sedimentary rocks		
Solid/labile rocks		Torrents in solid/labile rocks	Torrents in areas with large mass flow	
Unconsolidated rocks	Torrents in glacial valley fills (from natural reservoirs)	Torrents in recent moraines	Torrents in debris cones	Torrents in valley alluvials
	Torrents in other residual debris bodies			
Soils				Torrents caused by collapse of cliffs

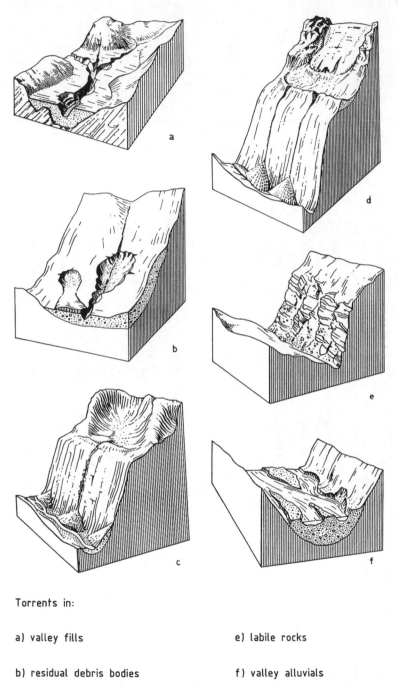

Torrents in:

a) valley fills

b) residual debris bodies

c) hard igneous and metamorphic rocks

d) recent moraines

e) labile rocks

f) valley alluvials

g) hard sedimentary rocks

h) debris cones

Fig. 6.10

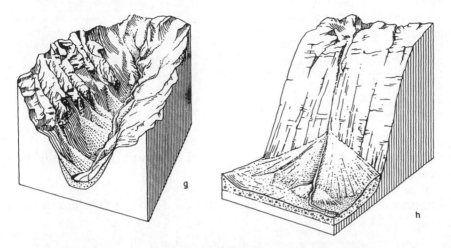

Fig. 6.10a–h. Types of Alpine torrents. (After Karl and Mangelsdorf 1975)

creeks and rivers is based upon the type of bedload. They differ mainly in that the solids in torrents are essentially of autochthonous derivation, whereas in mountain creeks they are mainly allochthonous in origin. Overhanging cliffs in mountain creeks and rivers are certainly also sources of solids, but the decisive point is the derivation of the majority of the solids. "It will, however, be necessary in the case of mountain creeks and rivers with large cliffs and long stretches of active downward erosion to talk of a local torrent character."

The highly variable morphology of torrents can be condensed into only a few basic types (Table 19, Fig. 6.10).

a) valley fills
b) residual debris bodies
c) resistant igneous rocks
d) recent moraines
e) labile-solid rocks
f) valley alluvials
g) resistant sedimentary rocks
h) debris cones.

6.3.2.2 Rivers of High Plateaus and Tablelands

These streams may be considered as a special type of mountain rivers (Keller 1962). From them Davis (1890) developed his system of river classification. In this type the rivers either adjust themselves to the topography and to the geological structure, or they form the topography in accordance with the alternations of the stratigraphy (Fluck 1925). Davis distinguished between:

– consequent rivers following the natural declivity of the land surface;
– subsequent rivers flowing into the consequent rivers from the sides at right angles to the dip and parallel to the strike;

C = consequent, S = subsequent, R = resequent
O = obsequent, I = insequent streams, L = original land surface

- resequent rivers as tributaries to the subsequent ones more or less parallel to the consequent main rivers;
- obsequent rivers flowing against the dip of the beds;
- insequent rivers, which show no clearly apparent relation to the dip of the beds.

These various types of rivers are schematically illustrated in Fig. 6.11.

6.3.2.3 Rivers in Low Mountains and Hilly Country

Certain aspects appear to favor setting apart the rivers of the low mountains on deeply eroded shield areas. They are rather low in bedload and suspended load. This also applies to the hilly country of originally glacially overprinted regions. The rivers here carry gritty or sandy material and frequently were reconstituted already several centuries ago by extensive cutting of forests, by mill dams, and irrigation systems. These well-established structures are completely integrated into the countryside and thus give the appearance of a natural development whose original character can be only vaguely recognized (cf. Bremer 1959; Bauer and Tille 1968). One could call such waters rivers of civilized regions, as they were in this condition already at the time when the major river regulations of the 19th century commenced (cf. Chap. 7).

6.3.2.4 Rivers of Plains, Mixed Types

As rivers of the plains one could consider smaller lowland streams as well as large rivers. They are rarely encountered in western and central Europe, as the influences of the high and low mountains are felt down to the mouths, climatically through the intensity of the discharge and especially morphologically through deposition and erosion. Keller (1962) counted amongst the rivers of the plains also streams such as the Niers and Kendel along the lower Rhine, which itself is not looked upon as a plains river. One can classify as belonging to this type, however, the large rivers of the Russian shield areas, including most of their tributaries, as well as many of the South American rivers. In the latter cases the influence of the wide plains completely dominates that of the source rivers in the mountains or in the hills of the Mato Grosso. In North America the rivers of the plains are less frequent than one would expect. This is due to the fact that in the northern parts the influence of the glaciation is still morphologically felt, and that a so-called fall line separates the old shield area from the coastal plains. This results in a nonequilibrated declivity.

6.3.2.5 River Mouths

The river mouth is a zone of particularly pronounced interaction between land and sea, between erosion and accumulation, where at one time the one force and at the other the other force will gain the upper hand. Two main types of river mouths have to be distinguished: firstly the junction of a tributary with its main stream, and secondly the entry of a river into a depression, i.e., into a lake or the sea when filled with water, or as an inland mouth or delta when the receiving area is dry.

Tributary/Main Stream. When a tributary is joining its main stream, not only the larger discharge and with this the larger transport capacity of the latter are of importance, but also the gradient and the bedload transport of the tributary (cf. Chap. 5.3.2). In the originally glaciated high mountain ranges there are numerous side valleys joining the main valleys with steep gradients or even hanging valleys

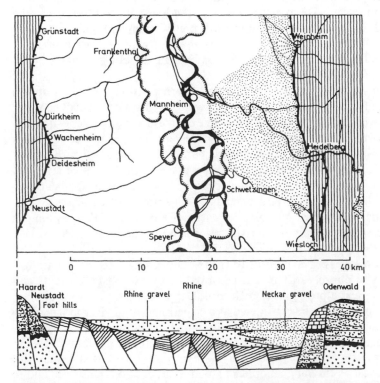

Fig. 6.12. Alluvial fan of Neckar near Heidelberg. (Redrawn after Wagner 1960)

which are drained by the respective tributary over a water fall or, in cases where erosion has already progressed somewhat, in a junction gorge. The debris cone which it builds into the main valley can be strong enough to displace even a large main stream to the opposite flank, leading to a shifting of the main river in its own valley. In some cases a tributary can join the main stream steeply without any development of a debris cone. A good example for a repeatedly displaced river is the inner-Alpine stretch of the Inn.

The bedload introduced by the tributary can lead to a temporary retention of bedload in the main river, which sooner or later will make itself felt downstream by increased transport rates.

In shallower areas the tributary will establish a debris cone which morphologically will not be very pronounced, but which can notably enrich the bedload spectrum and budget of the main stream. This is the case for the Danube and its Alpine tributaries Iller, Lech, Isar, and Inn (Bauer 1965, etc.). Reference must be made here to the spatial and thus temporal character of a debris cone (Wagner 1960, etc.), the sediments of which can mingle with those of the main stream over longer periods of time (Fig. 6.12).

When a tributary is only small or when it carries little bedload, a bedload-rich main stream may form an obstruction at the tributary's mouth through its supply of bedload, thereby forcing the tributary to flow on other alluvials over a longer distance, until eventually the tributary manages to cut across into the main stream. Such a displaced mouth is observed in the case of the Ill near Strassburg.

The smaller the grain size of the bedload or the lower its concentration, the smaller the break in gradient at the mouth will be. Whereas in higher mountain areas even a small tributary can influence its main stream at least over short periods of time, this situation is virtually unknown in the alluvial plains.

River Mouths in Lakes. The entry of bedload-bearing rivers into lakes always takes place in the form of a delta, only part of which will appear above the water level. Due to the lack of stronger currents, the coarse bedload is deposited immediately and forms rather steep slopes. Only the very fine-grained components are transported farther out and during floods can lead to increased turbidity of the lake water. During summer the colder river water will partly descend rapidly along a sharp boundary, leading to thermal layering. The suspended load can descend along with the water. This phenomenon is very pronounced in Lake Constance, but can be observed at all points of entry of rivers into lakes, as long as the content of suspended load in the river water is large enough (cf. Chap. 3.3).

River Mouths in the Seas. The mouths of rivers debouching into the sea are furthermore subject to entirely different conditions. Here the coastal currents and the tidal changes exert a pronounced influence. Depending on the point of view, the river mouths may either be considered as a phenomenon of river morphology or else of the marine sciences taking in all fluvial factors along the sea coasts as parts of the coastal morphology. There is justification for both points of view.

It has become common usage to treat the two main types of river mouths, deltas and estuaries, as fluvial phenomena, albeit in an area of transition, and to consider other phenomena such as coastal bars, limans, rias, as well as fjords and other types of "drowned" valleys as diverse aspects of coastal morphology. The latter can also be considered as a special type of valley (cf. Chap. 5.4.1).

The problems of bedload transport and the resulting formational processes at the river mouth are so complex that it has not yet been possible to establish a satisfactory comprehensive theory for the respective phenomena. We can do no more than interpret a few of them and to estimate their influence on the total system. Of special importance are:

- the incoming bedload yield and its relation to the discharge;
- the tidal differences in water level;
- the influence of the Coriolis force (cf. Chap. 3.2.3);
- the stratification at the meeting of fresh and sea water (cf. Chap. 3.3);
- marine influences such as coastal currents, surf, and wave action;
- the influence of prevailing winds.

A comprehensive discussion based on hydrodynamic principles was presented by Tison (1968).

Estuaries. This type of mouth is found along macro-tidal coasts where rivers with moderate quantities of bedload enter. An estuary remains open only where the transport of sediment is not sufficient to fill the funnel-shaped mouth, or where the tidal currents are strong enough for a continuous removal of the sediment. These currents form deeper channels amenable to navigation, but lead to the deposition of the material removed in the form of underwater bars farther out in the mouth. These are, for example, the bars or sands in the outer delta of the Elbe so much feared by sailors. The North Sea, the French Atlantic coast, and the Irish Sea are macro-tidal and their major rivers such as, for example, the Thames, Elbe, Weser, Seine, Loire, and Gironde essentially have funnel-shaped mouths. An exception are the Rhine and the Meuse with their common delta, as in the southern North Sea the tidal differences are smaller.

Deltas. Deltas are established where rivers rich in solids enter micro-tidal portions of the sea such as in the Mediterranean (Nile, Po, Rhône), the Baltic Sea (Vistula), the Black Sea (Danube, Kuban), the Caspian Sea (Volga), parts of the northern Arctic Ocean (Lena) etc., but also in the southern North Sea (Rhine-Meuse).

Wilhelmy (1972) gave a brief compilation of the preconditions for the establishment of a delta:

- There will be no delta without a large supply of solids.
- Greater depths, i.e., steeper gradients of the continental shelf are unfavorable for the formation of a delta.
- Strong coastal currents and large tidal differences inhibit or retard the formation of a delta.
- Along subsiding coasts, deltas are formed only where the amount of subsidence is compensated or exceeded by deposition of sediment.
- The growth of a delta will be strongest where microtidal intracontinental seas are bordered by tectonically inactive coastal plains.

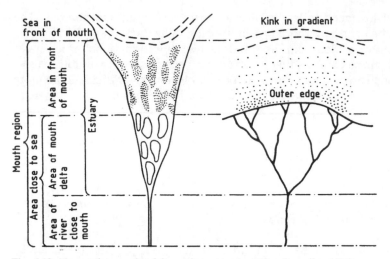

Fig. 6.13. Nomenclature of a delta and an estuary. (After Samojlov 1956)

It has been frequently pointed out in literature that morphogenetically, deltas develop by rearrangement of estuaries through numerous intermittent stages. This is possible where the conditions of sedimentation or the tidal variations change due to fluctuations of the sea level, but it should not be the rule. It cannot be excluded that during a rapid eustatic rise of the sea level the sediment supply temporarily will not suffice and that estuarine conditions will be established as a quasi-stationary stage. On the other hand, a rapid retreat of the sea level will result in increasing erosion, leading to a freshly incised river valley. An important factor is always an ample supply of sediment and low tidal differences. As a consequence of the eustatic fluctuations of the sea level in geologically recent periods, the up-river or down-river migration of deltas and estuaries can be observed, and at times a transition from one type to the other; this, however, is not the general rule.

Subdivision of Estuaries and Deltas. In the region of the mouth of a river a multiplicity of different areas may be discerned, from a stretch of pure flow to fully marine conditions. For the two types of mouths Samojlov (1956) proposed a schematic tripartite division (Fig. 6.13). Depending on the degree of predominance of marine or fluvial conditions he distinguished:

— the stretch of the river close to its mouth;
— the area of the mouth itself;
— the area in front of the mouth.

The first area starts where the lower reach of the river begins to widen and where during "normal water level" the influences of the tides or of the damming up of the river dissipate, i.e., where the effects of any fluctuations in sea level cease to be felt. In rivers with rapid flow this stretch may be missing; the influence of the sea is then restricted to the actual mouth of the river.

The area of the river mouth is the actual estuary or delta. It starts where the river splits up into several arms or in mouths having only one arm. There in the channel or in the funnel a delta-like structure develops with shallows and intervening channels. In the case of the estuary, this orificial area grades into the wide open area in front of it, still containing accretions of sediment, corresponding to the protuberances of the delta arms. For a delta this area ends along a conceptual curved line joining the mouths of the individual arms. In both cases these areas are called "coastal areas" by Samojlov.

The area in front of the mouth has an open border to the sea. If there is a break in gradient between the submarine part of the delta in front of the arms or between the estuary funnel and the sea floor, the outer orificial limit will extend to this line. Where there is no such break in gradient as a boundary, the zone to which the influence of the flood stage of the river will approximately extend can be used. This area is no longer of interest from the point of view of river morphology, belonging rather to the field of coastal morphology.

The generally accepted definition of a delta locates its base at the point where the river splits up at a break in gradient into arms, and the development of an actual sedimentary cone takes place. A further tripartite subdivision of the delta itself into an inner part near the initial split-up, a central part, and a part close to the sea, although acceptable, is not really warranted. The investigation of deltas has grown into a separate branch of hydrological science. Its extensive literature can only be referred to here cursorily. A more detailed discussion would considerably exceed the scope of the river morphology.

6.3.3 Climate-Controlled Types of Rivers

This heading covers the multitude of groups described under Section 6.1 above, i.e., perennial, episodic, and periodic streams on the one hand and the endoreic and diareic on the other, as well as autochthonous and allochthonous ones. A few other terms will be discussed below.

Rivers which are intermittent, in that they regularly fall dry, are covered by the Italian expression "fiumare". These are all streams which have to rely on the direct supply of water from the precipitation and which are not sustained during the dry summer season by the inflow of groundwater. As indicated by the derivation of the term, this type is widespread around the Mediterranean.

The term "torrente" (Italian, Spanish) is used differently. On the one hand, it is used for a true torrent, as in France, on the other hand it is synonymous for the fiumare, especially in the southern Alps (Hormann 1964). Adhering to the French definition, this author refers to torrents in the wider sense as meaning "channels with highly variable discharge and ample supply of gravel". Torrents in the strict sense would then be the fiumare which, however, the author does not refer to as such. Hormann divided the torrents themselves into two groups, those of the circum-Mediterranean area (the actual fiumare) and those of the humid regions for which bedload transport is of prime importance and where the discharge disappears in the residual debris ("debris cone torrents"). The definition

of torrents is rather confusing and is not applied in river morphology. The term
"torrente" is used in official geographical designations such as Torrente Cellina
which, however, need not conform with the scientific definition of torrents (Hormann).

In Australia the term "creek" is used for dry valleys which locally carry some
water and can widen into strings of lakes. Their function is similar to that of the
wadis of the Sahara and the Arab world. In South Africa they are referred to as
"riviers".

6.3.4 Discharge Regimes

6.3.4.1 Definition

Averaged over a longer period of time, every perennial river will show a certain
discharge hydrograph which will be the direct result of the geographic features of
its drainage area. These features are so numerous that no discharge hydrograph
is similar to the next. Nevertheless, during observation and comparison of the
discharges from different river areas common features frequently evolve, leading
to the subdivision of rivers into hydrological types, the basis of which are the so-
called discharge regimes. A regime is defined by the mean annual discharge curve
and the main hydrological parameters, previously known as the main "numbers".
As the shape of the discharge curve is of greater importance than its absolute
magnitude, it has become customary to relate the monthly averages to the annual
mean, taken as 1.0, in order to achieve a better comparison. Lately, for the
description of the regime type, statistical methods have been employed, amongst
them notably the Fourier analysis (Keller et al. 1972).

All factors controlling the annual trend of the discharge are referred to as
regime factors. In addition to the well-known components of the global circulation of water, i.e., precipitation, evaporation, storage, and consumption, great importance must be assigned to the morphological factors of a drainage area, the
condition of the bedrock, and soil cover. Specifically, the precipitation is of controlling influence. However, its annual trend can be superimposed by other regime
factors to such an extent that it can no longer be recognized in the discharge. Such
a factor can be fixation of large amounts of precipitation by snow cover and
glaciers.

That there is an influence of the morphometry of a river system on the discharge regime is clearly evident. There are also indications as to the opposite, i.e.,
that in the interaction of natural forces the regime also participates in the shaping
of the river and valley geometry. As an example, reference can be made to late-
or post-glacial terraces along the rivers of the Alpine foreland of Bavaria, which
owe their origin to the previously much higher discharge having a very pronounced summer maximum. The discharge regime presently prevailing in these
regions would not suffice to explain these morphological features. The same applies to the ancient river valleys originally running parallel to the edge of the inland ice in Europe and North America, as well as to the adjoining periglacial

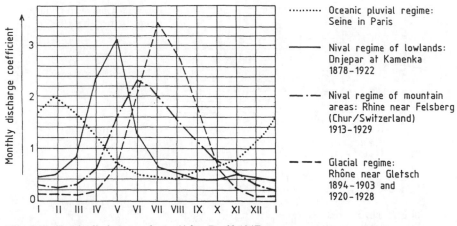

Fig. 6.14. Simple discharge regimes. (After Pardé 1947)

areas. For an understanding of the varied morphology of river country, it will be helpful to consider also regime types, even when their relation to river morphology is only rather loose. In order to avoid misunderstandings, it has to be stressed that the concept of discharge regimes has no connection whatsoever with the so-called regime theory (Chap. 5.6).

The best-known classification of discharge regimes was given by Pardé (1947, etc.). The following overview is based on the compilations by Keller (1962) and Marcinek (1975).

6.3.4.2 Discharge Regimes after Pardé

According to Pardé, the discharge regimes are subdivided into three main groups with a number of subgroups.

1. Simple regimes. Simple regimes possess only two hydrological seasons, i.e., a flood- and a low-water season. There are relatively large fluctuations between the two levels which are somewhat reduced only in oceanic pluvial regimes (Fig. 6.14). Simple regimes belong to a climatically uniform area. They are subdivided into:

1a) Glacial regime. This is established when at least 15–20% of the drainage basin is covered by glaciers. Storage and consumption of snow and ice mask the precipitation to such an extent that it is barely recognizable in the discharge curve.

1b) Oceanic pluvial regime. The winter maximum of precipitation, which in oceanic climates falls only to a very small extent as snow, together with the large demand for water during the summer evaporation leads, in contrast to the glacial regime, to a discharge maximum from January to March, and a minimum during late summer.

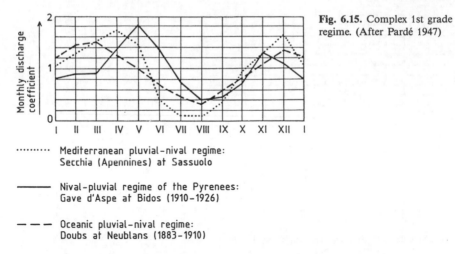

Fig. 6.15. Complex 1st grade regime. (After Pardé 1947)

·········· Mediterranean pluvial-nival regime:
Secchia (Apennines) at Sassuolo

———— Nival-pluvial regime of the Pyrenees:
Gave d'Aspe at Bidos (1910-1926)

— — — Oceanic pluvial-nival regime:
Doubs at Neublans (1883-1910)

1c) Tropical pluvial regime. The discharge curve of tropical rivers is controlled by changes between rainy and dry seasons following the position of the sun. Thus no uniform dates for floods and low waters can be given. In the northern hemisphere the maxima are between July and September, in the southern hemisphere from February to April. The respective minima are shifted by 6 months each.

1d) Nival regimes of mountain areas. This is characterized by a winter minimum in discharge and a maximum in July or later, depending on the advance of the snow melt to higher elevations. If glaciers occupy a large proportion of the drainage area, the discharge in July and August can be higher than that in May. The regime then is called nivo-glacial.

1e) Nival regime of the lowlands. This governs the rivers of the continental plains regularly covered by snow at low temperatures in winter. The rapid rise in temperature during spring will result in melt water floods, at times increased by rain fall. In contrast to the mountains where the melting of the snow cover gradually proceeds to higher elevations, it sets in rather abruptly in the lowlands and thereby leads to particularly devastating floods.

2. Complex 1st grade regimes. This group covers rivers, the regime of which is governed by several factors such as ice, snow, and rain. They frequently show two, at times even three maxima. Compared to the simple regimes, their discharge is more uniform as several sources of supply are available. The ratio between the monthly means and the annual mean rarely rise above 2 (Fig. 6.15). The most important examples for this type are:

2a) Nival-transition type. The first maximum caused by the snow melt in June is followed by a second, caused by the winter rains in November or December. Throughout the summer there is little change in the discharge.

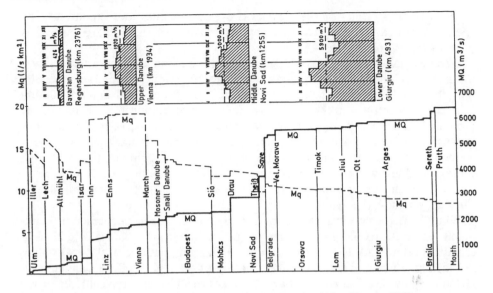

Fig. 6.16. Discharge regime and mean discharge of the Danube. (After Kresser 1973)

2 b) Nival-pluvial type. This regime has two pronounced maxima and minima. Depending on elevation and snow cover, the main peak may be observed in April to May, the secondary maximum in November to December. In the Mediterranean climate, with autumn and winter rains, both maxima can attain the same size.

2 c) Pluvial-nival type. In this regime the contribution of the snow melt to the total discharge is of subordinate importance. The main maximum in April, being increased by the snow melt, usually slightly exceeds the November maximum, which is caused entirely by rain. The summer evaporation of this regime type, which is mainly encountered at elevations of 1000−1800 m, leads to the summer minimum falling even below the winter minimum.

2 d) Types with more than two maxima. Regimes with more than two peaks of discharge can be observed in northwestern Japan. They are caused by the retardation of the discharge due to retention in the snow cover, and the monsoon circulation outside the tropics. In the western part of the Suebian section of the foreland of the Alps, three maxima in the discharge curve are developed. The first maximum in March is caused by rain, augmented by the snow melt in the Alps, the second one, in July, by the summer precipitation, and the third one, in November, by the oceanic winter rains.

3. Complex 3rd grade-regimes. This group covers all rivers in the discharge area of which several different discharge regimes are developed concurrently. This applies to most of the larger rivers which on their course are crossing different discharge regimes such as the Rhine, Rhône, Po, and Danube. Under these regimes are also included those rivers which on the one hand belong to the pluvial type, but which cross different climatic zones. Examples are the Nile, Niger, and Amazon.

As a final example for the complex 3rd-grade regimes, the Danube will be briefly described. A comprehensive characterization of its discharge regime is given in Fig. 6.16.

Whereas in its upper section down to Ulm, the Danube is characterized by a spring maximum and a low summer discharge, the Alpine tributaries gradually shift this maximum into the summer. At the same time, the tributaries joining from the hilly country to the north with their early rises in water level due to the snow melt lead to a dampening of the annual discharge curve of the Danube. Downstream from Passau, the Inn, more than any other tributary, leaves its mark on regime of the Danube and transforms it into a typical mountain river with high summer and low winter discharge. Even the tributaries joining the Danube on its course through Austria and Hungary are not able to change this picture basically. They only reduce the extremes.

Only in Yugoslavia is a transformation to another discharge characteristic initiated, which the river then maintains down to its mouth. The Drava with its rather uniform flow leads to a rise of the winter minimum, but also of the peak values in May and June. The decisive transformation of the regime of the Danube is effected only by the Tisa and the Sava. Whereas the Tisa as a typical river of the plains has its maximum in spring and its minimum in autumn, the Sava is controlled by Mediterranean influences which lead to a dry summer and high precipitation from November to May. From the mouth of the Sava onward, the time of regular floods falls into the period of August to October. The tributaries below the Iron Gates, from the Carpathians and the Balkans, at times show dangerous floods, but they are not able to cause decisive changes of the discharge regime of the Danube.

7 River History

7.1 River History as Part of Geological History

The principle of actualism, "the most important conceptual basis for the interpretation of geological phenomena", states "that forces and processes of the geological past are similar to those observable today, allowing direct conclusions from the phenomenon observed to the initial processes" (Murawski 1983). This applies to all important parameters, with the restriction, however, that during the course of geological history several peak phases of the usual processes such as folding, volcanism, and glaciations occurred. As a consequence, the present picture does not always permit a satisfactory interpretation of these processes. Despite the acceptance of the principle of actualism and the observation of considerable periods of geological history, the presence of certain peculiarities has to be expected.

Geological phenomena, apart from paleontology, can be traced back by the investigation of

- the geomorphology,
- the rocks themselves.

In the case of streams these are erosional forms such as valleys and terraces and accumulative forms such as valley fills, debris cones, and deltas, as well as limnofluvial sediments.

Remnants of land forms can be traced back with some certainty to the upper Tertiary only. Farther back into geological history one has to rely on the analysis of the rocks and their stratification, as well as on the preserved configuration and extent of lithological units. In marine and terrestrial sediments directions of transport are recognizable which allow conclusions as to their source areas. The sediments themselves also furnish information on the depositional facies. By its fluvial sediments an individual river will not be recognizable as such, even when locally the channel-like character of the deposit becomes quite distinct. What can be observed is the transport mechanism and its main directions which are entered into paleogeographic maps.

The body of rock is a three-dimensional feature, the formation of which required hundreds of thousands or even millions of years. Nature is able to sustain sedimentary cycles over considerable periods, established directions being used time and again. The individual river here cannot be specified, but only the reaction of the earth's surface to the endogene forces and the prevailing climate, the sedimentary body representing, as it were, the sum of this interaction. This lithological unit originally showed a multitude of structures which later were largely obliterated by tectonics (diagenesis and metamorphism). Frequently old fluvial gravels and sands have been and are used again as bedload material which eventually can become a marine sediment with its fine-grained components.

The major manifestations of terrestrial sedimentation such as the Triassic Buntsandstein of Europe, apart from its desert formations, have been built up by innumerable smaller and larger rivers. The resulting rocks, however, cannot be considered from the point of view of fluvial history, as there is no individual river identifiable as a potential target of research. River history here has naturally played a prominent role, but only few traces can be reconstructed. This will also be the case in the distant future with our present rivers. Nevertheless, each river presents itself as a three-dimensional, individually shaped unit with its own course of development. Its character as an interwoven system allows conclusions from its past development as to its future course. The reactions of a river to certain regulating measures can also be estimated much better, the more we know about its history.

For a keen observer of river history this tends to emerge only during the uppermost Tertiary. It is an important object to Quarternary research and comes fully into its own with the retreat of the inland ice since the end of the last glaciation. Climatic developments and tectonic processes each imprint their own individual patterns and land forms. A strict division by the one or other agency is not always possible, and in the discussion below the respective predominant factor will be stressed.

7.1.1   River and Climate – Exodynamics

7.1.1.1   Formation of Terraces

The continuous change between warm and cold periods during the Quarternary has resulted in a larger number of distinctly subdivided regions of terraces, or meander valleys with all their stages of transition (Chap. 4.5.2.4) which locally are very difficult to interpret. From the host of publications on this subject, a number of references will be quoted on the nomenclature for terraces and its application.

In general the cause of the development of river terraces lies in the continuous, climatically controlled change between lateral and vertical erosion. These are related to advance, standstill, and retreat of the ice margins, as well as to tectonic factors such as ascent and subsidence of the crust (cf. Sect. 7.1.2). This results in three problems:

- with growing distance from the ice margin a distinction is made difficult, especially in the periglacial areas which had not been covered by ice, as here only an indirect comparison is possible;
- later erosion will lead not only to a modification of the older terraces by the younger ones but even to a complete destruction of the former;
- the older the terraces are, the more difficult they will be to classify.

Several types of terraces have to be distinguished:

1. According to the type of setting. Depending on their type of setting, rock terraces and terraces in unconsolidated sediments are considered separately. So far, however, the terminology applied is far from clear, as the various terms can be applied with different meanings. Weber (1967) stated that "with regard to the nomenclature of river terraces there is considerable confusion . . .". The textbook of Kayser (1923 in Weber 1967) calls erosion or rock terraces the old gravel-covered valley floors cut from a 'solid' bedrock. Frequently the 'rock terrace' in the modern sense, i.e., the gravel-free shallow slope, is considered the result of lateral erosion which supposedly was not associated with sedimentation. The accumulative terraces according to Kayser's textbook are erosion terraces cut into gravel beds and not into the pre-fluvial bedrock. Philippson, on the other hand, called "depositional or accumulation terrace", a sedimentary body formed only after the widening of the valley by supposedly nonsedimenting lateral erosion. Such terms give little satisfaction as none of them clearly states what it is supposed to imply.

One reason for this situation is that bi-lateral erosion and vertical erosion can take place simultaneously, thus leading to the formation of a river bed. Furthermore, lateral erosion can be accompanied by deposition, but not necessarily so. Machatschek (1953) and Troll (1957) stressed the importance of these processes in climates with a sudden high discharge. The "stepping-down" of a stream is considered by v. Wissmann (1951) as a climatic morphological phenomenon.

2. According to their time of formation. The conspicuous differences between the terraces with regard to position and setting in rocks as gravel terraces invited another distinction based on the time of formation, especially since Penck and Brückner (1909) were able to recognize the sequence of the periods of glaciation and their spatial arrangement.

By this time criterion again two types of terraces can be distinguished. These are firstly terraces which yield evidence for a complete period of glaciation (Fig. 7.1) i.e. for a climatic megacycle, and secondly the numerous late- to post-glacial features after the last Würm/Vistula (Wisconsin) glaciation giving witness of

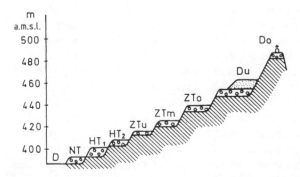

Fig. 7.1. Section through a terraced valley. (After Graul in Weber 1967)

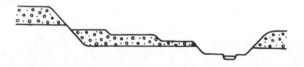

Fig. 7.2. Younger gravel terraces

short climatic fluctuations (Fig. 7.2). It can be assumed with certainty that such features were also developed during the earlier glaciations, but they vanished during the subsequent periods. They either completely disappeared or they were reworked and modified to such an extent that they can no longer be distinguished. However, there is still some uncertainty as to whether each older terrace stands for a long period of glaciation. Despite the progress achieved in such research, the timing of terraces is still beset with uncertainties, as fossils or paleosols and other witnesses of the climate which would facilitate their relative dating are rare.

It is, however, evident that an inversion of topographical relief has occurred, i.e., the oldest terraces are situated at the highest and the younger ones at the lowest elevations. Each glaciation with all its phases has cut itself into and overprinted the preceding valley formation. The Quarternary time scale in terraced country proceeds from the top downward, from the highest peneplanes down to the present-day river. This tendency can be traced back to the uppermost Pliocene. Intervening phases of deposition were not able to stop this "stepping-down" of the rivers, which has been described as a rhythmic process which only made possible the formation of river terraces. A similar sequence in principle also applies to the late- and post-glacial terraces (Troll 1954; Diez 1968).

The eroded older sedimentary material ends up in the lowlands or the sea. The Hungarian basin, an area with a long tendency of subsidence, for example, contains more than 800 m of upper Tertiary and Quarternary sediments supplied by the rivers Danube, Raab, Drava, and Sava, coming essentially from the Alps, and by the Tisa and its tributaries from the Carpathian mountains. In such basins the stratigraphic sequence is normal, i.e., the oldest rocks are at the base, the youngest on top.

7.1.1.2 Change of Land Forms in Coastal Areas

During the Quarternary special conditions prevailed along the coastlines of the oceans. The repeated immobilization of huge volumes of water by the inland ice during the periods of glaciation led to various phases of retreat of the sea level and to a rise during the subsequent warmer interglacial periods. This is generally referred to as eustatic fluctuations of the sea level. Hence vast beach terraces were formed which presently lie either above or below sea level (e.g., Wagner 1960; Woldstedt 1969; Schwarzbach 1977). Whereas the Quarternary stratigraphy north of the Alps is established with the aid of valley or river terraces, locally in connection with moraines, this can be done in the Mediterranean region with the aid of beach terraces or "raised beaches". The sequence here is also from the top downward, with the same problems of the terraces as witnesses for longer cycles and of the relatively short stages postdating the last glaciation. The rivers would naturally react to these changes in sea level. During phases of retreat their courses

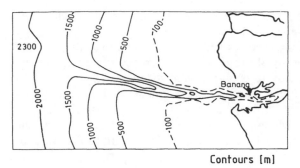

Fig. 7.3. Apparent direct continuation of a river mouth on the continental shelf. (Zaïre River − Atlantic)

Contours [m]

were extended, to be shortened during an advance which drowned parts of their lower reaches. These phases were separated by periods of thousands or tens of thousands of years. River mouths would have migrated several times during the Quarternary (cf. Chap. 6.3.2.5).

Depending on the supply of sediment, the drowned lower reaches may have been filled to various degrees. Locally they can still be traced, however, along the sea floor. Additionally, deep channels are developed on the shelf regions which are referred to as submarine canyons. They can attain astonishing dimensions in length and depth, and exhibit great similarity to drowned river valleys. They show only indirect connections to modern river courses, if at all. Such apparent submarine continuations of a modern river are shown well developed in front of the mouths of rivers such as the Hudson, Zaire (Fig. 7.3), Indus, and Ganges/Brahmaputra. There are, however, several hundreds of such channels quite detatched from any present or ancient river. Intensive research has excluded fluvial erosion as well as tectonic processes as likely causes. Most probably they have been formed by submarine density currents of elevated specific gravity being due to larger concentrations of suspended load. As "turbidity currents" of considerable extent they are able to exert correspondingly large erosive effects (Daly 1936; Kuenen 1953; Dietrich 1959). Once established, such channels were used time and again and thus were deepened further. Density currents have been referred to under Chapter 3.3. The transportation capacity of these turbidity currents along the edge of the continental shelf is considerable, as evidenced by the large dimensions of the canyons. They are up to 100 km long and some 100 to 1000 m deep, frequently branching out, and can be traced down to a depth of 3000 m.

Such canyons are a rather recent feature geologically, locally cutting into upper Tertiary strata and even solid bedrock. It cannot be excluded that their formation was initiated mainly during the Pleistocene, when the extended rivers, following a retreating sea level, dumped large amounts of sediment onto the shelf. Simply due to their own weight these could have caused turbidity currents which were then continuously repeated. This would explain at least the small number of canyons in front of the mouths of rivers both ancient and modern.

The vast erosive and depositional effects exerted only by rivers postdating the last glaciation are especially notable through human activities. These have left benchmarks in time and space which allow the recording of measurable changes over defined periods of time. These are most pronounced in the deltas, with their slow advance with time. Not only did a large number of late- or postglacial lakes disappear, but large marine inlets were likewise silted up.

The seaward advance of a delta is controlled by the supply of sediment, while the depth of the sea floor naturally also exerts a considerable influence. The extremes vary between a shallow bay receiving a large sediment-rich river, and a deep basin with river mouths supplying only little sediment. Deltas can even account for a large part of the total drainage area of a river: Terek 16%, Ganges-Brahmaputra 4.8%, Volga and Mississippi 1.1%, Nile 0.8%. The extent of the silting can be illustrated by a number of examples drawn mainly from Wagner (1960).

During the Pliocene the Rhône valley terminated near Lyon, forming a gulf which has been filled since by the Rhône and its tributaries. The sediments of the present delta are up to 300 m in thickness. Around 500 B.C. the town of Arles was reported to lie at the coast, and is now situated 50 km inland. To mark the main channels of navigation it was customary to erect pairs of lighthouses. Five to seven pairs, which had to be abandoned one after the other, are found inland. The last tower was built in 1793, and now lies 10 km inland. The present growth rate of the Rhône delta is about 23 ha annually.

Pisa in northern central Italy was founded around 1000 B.C. as a harbor at the mouth of the Arno. This river is known for its high sediment discharge derived from the flysch portion of the Appennines. At the time of Augustus it was situated 4 km from the coast, the present distance being 12 km. Ostia, the oldest port of supply for ancient Rome, has a similar history of redundancy. Already during Imperial times it required a second adjacent location which also was eventually silted up. The remnants of this settlement are now about 5 km from the coast.

Parts of the Greek and Anatolian coasts also suffered major changes. In the Gulf of Thessalonica, the ancient Thermae, after the 5th century B.C., some 800 km^2 were silted up by the sediments of the Wardar (Axios) and transformed into the Macedonian Campagna. The harbor existing since the Classic period is situated about 20 km away to the side of the mouth of the Wardar and was thus able to "survive".

The Anatolian rivers Hermos, Kaystros, and Maiandros, to name just the best-known ones, have led to the decline of numerous towns and trading ports. The ports of places such as Heracleia, Pyrrha, Priëne, Myos, and especially Miletos were silted up by the advance of the delta of the Maiandros. The port of Myos around 500 B.C. once harbored 200 ships, but already in early Roman times the town was abandoned, as the area had become a swamp. Strabo reported that Myos was situated 30 stadia (5.6 km) from the sea, whereas the present distance is 26 km. Miletos is presently about 7 km inland. The island of Lade, renowned as the place of a sea battle during the Persian wars, today is just a hill in the delta of the Maiandros, or Menderes as it is called now. The river Kaystros advanced its delta to such an extent that the town of Ephesos now is situated 5 km inland. Already during the Classical period the port had to be connected to the sea by means of a canal.

That Egypt is "a present of the Nile" was already known at the times of the Old Empire. Herodotos furnished a detailed description of the situation. The present-day delta measures some 22000 km^2.

Most likely the Euphrates and Tigris forming Mesopotamia established their common delta, the Shatt-el-Arab, only during historic times. Here we have the case of an expansive delta invading a shallow part of the sea, the Persian Gulf, silting it up for several 100 km. Marine Quarternary beds are traceable up to 400 km inland. Since about 500 B.C. the delta has advanced by 64 km.

Similar examples are presented by the Hwang-ho and the Yangtze-kiang of China, albeit with considerably larger deltas. The Hwang-ho delta extends over about 250000 km². It is advancing into the shallow Yellow Sea which it is estimated will become completely filled within about 25000 years.

The sediment-rich Colorado River has dammed off the northern part of the Gulf of California. The resulting sea water-filled lake shrank in the arid climate to form an inland salt lake, the Salton Sea, well below the present sea level.

The delta of the Mississippi has been the subject of intensive research, its history having become unraveled by many thousands of drill holes for oil and gas. The thickness of its sediments is wellnigh astounding, figures over 1000 m having been reported. This appears to be caused by the continuous subsidence of this part of the Gulf of Mexico, a tendency supported by the weight of the river sediments. On the other hand, we here note repercussions on the further development of the delta. Presently an equilibrium of the sedimentation prevails only in front of the most active arms of the delta, whereas shallow lakes are forming within the delta itself. However, the development of these lakes is far from uniform. It is suggested that the largest of these, Lake Pontchartrin, is underlain by a part of the crust subsiding below sea level.

Each delta has its own particular history which may be rather varied. From the point of view of geological history, all modern deltas are short-lived features, advancing only until the next eustatic sea level rise puts an end to them and leads to a restructuring.

The lower reaches of the Po bear description as an informative example for the dynamics of a river mouth. During the Würm glaciation the sea level was lower by about 95−100 m and the river extended nearly to the central part of the Adriatic Sea. With the post-glacial rise of the sea level which started about 10000 years ago, the sea encroached far into the Lombardy plain. Due to their high rate of solids discharge, the tributaries accumulated thick sedimentary deposits, measuring about 250 m at Modena, 90 m at Reggio, and 170 m at Venice.

The historic Po delta has by now been drained almost completely by nature and centuries of human effort. Now it has started to develop farther south than the present one, approximately between the Po di Venezia and Ravenna, essentially underlying what are now known as the Ferrara plains. The peculiarity and versatility of this river-made landscape already roused the interest of the ancient geographers. Polybios and Strabon devoted much time to a description of the complicated system of branches in the delta of the Po. The most detailed description of the river, however, was given by Pliny the elder (Nat. hist. III, 120). His reports, in connection with the recent contour lines and aerial photos of the Ferrara plains, allow us to locate the ancient mouths rather accurately. During Etruscan times the plains were still submerged by a wide lagoon, subdivided mainly in east-west direction by ridges along the then active channels. The various branches of the delta in fact led to the formation of levees standing out above the

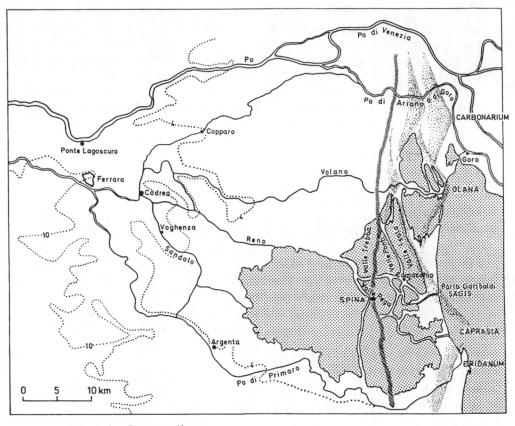

▨▨▨▨ Lido during Etruscan times

▧▧▧ Delta spits during Roman and Medieval times

⋯⋯⋯ Contour lines

Fig. 7.4. Reconstruction of the ancient Po delta. (After Alfieri and Arias 1958)

lagoon. In this way it is possible to trace a water course which has disappeared long since. From this reconstruction can be located the apex of the ancient delta close to the present town of Ferrara (Fig. 7.4).

Of the various contour lines, that at 4 m a.m.s.l. appears to give the most reliable information as to the location of the spits rising out of the lagoon already during pre-Roman times. The delta was closed off by a fairly straight barrier, the Lido, through which the branches of the Po had to cut their way to the sea. The former coastline is marked by the remnants of a magnificent line of dunes running about 10 km inland between the Po di Venezia and Ravenna. The port of Spina, important in pre-Roman times, was discovered along this Etruscan Lido southwest of Comacchio during drainage construction in 1924. The fact that for this settlement not one of the more sheltered sand banks was used, but the passage of the Sagris through the dunes speaks for the enterprise of this Greek-Etruscan community.

The political unification of the Po plains under the Etruscans established the preconditions allowing Spina to become the most important port for the whole hinterland. But with the collapse of Etruscan power the fate of this once florishing town was also sealed. The silting up of the lagoon and the subsequent subsidence of the coastal strip by about 2 m, both made Spina fall into oblivion (Alfieri and Arias 1958; Cati 1981).

In 1154 A.D. the Po breached its banks at the mouth of the Panaro near Ficarolo and opened new access to the Adriatic Sea north of the old delta. Initially this outlet advanced only slowly, but the annual rate increased after the construction of dykes had been started and the Po was no longer able to leave its bed as frequently as before. From 1600 to 1830 the area of the delta increased on average by about 135 ha a^{-1}, until 1893 by 76 ha a^{-1}, and since the turn of the century only by about 50 ha a^{-1}. This decrease is caused by the advance into increasingly deeper portions of the Adriatic Sea. The apex of the delta now is at Serravalle where the river splits up into the Po Grande and the Po di Goro. From Mesola, which 120 years ago was at the coast, the delta now has advanced some 20 km farther to the east.

The discharge of the various arms of the delta has changed repeatedly. Since 1872 the Po della Pila, running due east as the main arm, bears most of the discharge and consequently the most extensive sand banks are formed ahead of its mouth.

The following general rules thus apply to the formation of a delta:

The sand banks connected by a semi-circular barrier grow both vertically and laterally due to the surf. When a large flood event coincides with a storm period, the seaward barrier can diminish to such an extent that the river may force its way to the sea by a new arm. On the seaward bank of the arm, deposition of material from the river and the sea leads to the formation of a small coastal island. On the opposite bank a sand bank forms only by the river sediments. When the sand bank has grown sufficiently, a lagoon is formed and cut off from the open sea. In this way a network of delta lakes, the so-called valli is formed which silt up only slowly as the rivers flowing beyond the sand ridges bounding them take their sediment load mainly out to the sea. This explains the rapid advance of the delta and its cellular structure.

The sediment deposition in the sea so far has not exceeded the 30 m contour. From the topography of the sea floor, and based on the assumption that the coastline originally described a uniform circle from the Lido of Venice to the Comacchio lagoon, a total amount of 3×10^9 m^3 of sediments can be calculated. At an average sediment yield of 1.2×10^7 m^3 a^{-1} by the Po, the age of the present delta covering some 140 km^2 is obtained as just under 300 years. The true age will be somewhat greater, as a part of the sediments originally deposited will have been carried away by surf and marine currents (Martinelli, in Samojlov 1956).

7.1.2 River and Tectonics – Endodynamics

7.1.2.1 Formation of Terraces

Movements of the earth's crust may engender the same terrace morphologies as
do the climatic factors. Ascent leads to an increase in vertical erosion or is its actual prime cause. Subsidence decreases erosion, leading to its cessation and eventually to deposition. The tectonic movements can be restricted to small portions
of a catchment area or extend across the whole of it. Locally restricted subsidence
can also be caused by the solution of bedrock, such as in areas of karst weathering
or over salt deposits. The mobility of the earth's crust may be seen as prime motor
for the activity of rivers. Once initiated, it will continually revive them.

Local as well as widespread processes of whatever origin exert their influence
on terrace formation. Weber (1967) presented an example for this. In Fig. 7.5 a
river is assumed to flow from right to left. Portion S has been lowered due to
crustal subsidence after deposition of the youngest terraces so that valley floor
2 has been brought to the elevation of floor 1. In this case the relative height of
2 within the subsided section gives the erroneous impression of less antiquity. In
such areas of subsidence, in certain instances the younger valley floors may attain
a higher elevation than the older ones.

Further to be taken into account is that in transverse valleys, at places pronounced deflections of terraces may be encountered, i.e., the distance between the
valley floor and the upper edge of the terrace increases toward the areas of ascent.
This is referred to as a divergence of terraces (Fig. 7.6). Toward the edge of the
area of ascent, the differences in elevation between individual terraces decrease.
This is referred to as convergence of terraces. This applies to the climatically con-

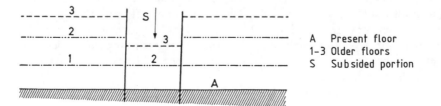

A   Present floor
1-3 Older floors
S   Subsided portion

Fig. 7.5. Modification of valley floor as the result of crustal movements. (After Weber 1967)

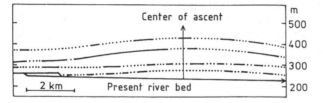

Fig. 7.6. Schematic section of tectonically warped terraces in an antecedent cross-cutting valley. (After
Wilhelmy 1972)

trolled terraces subjected to the ascent, as well as to the "tectonic terraces" caused directly by the ascent. Over labile portions of the crust where endogene and exogene movements are continually interacting, disturbances in level of older river sediments have to be expected everywhere. The accuracy of the topographical reconstruction thus depends on the type and size of the remnants available and on their elevation above the present river.

7.1.2.2 Transverse Valleys, Antecedence, Epigenesis, River Capture

Transverse valleys are a particularly conspicuous morphological feature. The term is not quite appropriate, but becomes understandable when the situation in question is viewed without prejudice. The rivers appear to "cut across" certain ground

a

b

c

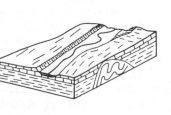

d

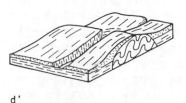

a'

b'

c'

d'

Transverse valley resulting from:

a–a' Overflow channel of lake

b–b' Headward erosion

c–c' Antecedent erosion

d–d' Epigenetic erosion

Fig. 7.7a–d. Origin of cross-cutting valleys. (After Wilhelmy 1972)

elevations and even complete mountain ranges, instead of flowing around these obstacles, as one would expect. Such valleys are clearly visible and measurable proof of crustal movements which can be active even today. Depending on their mode of formation through crustal ascent, regressive erosion, and change of bedrock lithology, four basic types of transverse valleys can be distinguished, with a number of transitional types and types of multiple origin (Fig. 7.7).

Antecedent Transverse Valleys. The river was in existence prior to the beginning of the ascent of the area concerned. The main condition to be met is that the erosive potential of the river is sufficient to compensate for the ascent. If it is smaller, the river will be deflected by the slowly rising obstacle. If erosion by the river, however, is stronger than the ascent rate, or when the latter proceeds without major interruption, the formation of terraces may fail due to the strong erosion.

Antecedence is rather frequently developed. One of the best-known examples in the Rhine cutting across the Rhenanian Slate Mountains in a world-famous scenic valley. The last ascent which gave the area its present morphology has been continuing since the upper Tertiary. The Rhine and its tributaries have incised in stages and left a whole series of terraces. This downcutting continued throughout the whole of the Quarternary. Other examples of this type of valley are that of the Elbe through the Elbsandstein Mountains in eastern Germany and of the Danube through the Iron Gates along which the active fold belt of the Carpathian Crescent was surmounted.

For many of the valleys cutting across the tectonic grain of the Alps, antecedence also has to be assumed which had been outlined since the Tertiary. However, depressions of the longitudinal tectonic axes of the mountain range were frequently used here as a starting point, such as is the case for the Alpine Rhine and the Inn.

A conspicuous example for antecedence is given by the Green River of Wyoming and Utah in the USA, which cuts across the Uinta Mountains, not in a straight line but with several attempted detours. Such rivers composed of several transverse reaches and detours are frequently developed and are present, for example, in the Indus, Brahmaputra, Euphrates, and Ebro systems.

Epigenetic Transverse Valleys (Change of Lithology). These valleys are formed where a river cutting into its bedrock suddenly has to cut across with an increased gradient as there is no possibility for a deviation. After a subsequent removal of the softer cover and the sculpturing of the more resistant escarpment, the impression may arise of a river reacting contrary to all established rules.

It would apparently have been easier for the river to flow around this obstacle. The epigenetic transverse valley is rather narrow and, due to the increased resistance of the rock, also steep-sided. Epigenesis is frequently developed in tablelands. "Just as the Danube at Geisingen, such rivers directly approach the front escarpment of a rock layer and cut across it in a gorge or at least a narrow valley. From one point of view these rivers are not in disagreement with a bedrock: they are flowing more or less parallel to the dip of the strata. Initially all rivers were consequent, i.e., they followed the declivity of the freshly emerging land surfaces. The relief ... followed the stratification of the beds deposited at least prior to the emergence of the sea floor. With progressively wearing down the original surface, the differences in lithology assumed increasing importance in moulding the landscape. This led to different conditions of declivity, compared to which only the larger, more erosive rivers could maintain their direction. Streams with less discharge had to adapt to the new situation brought about by denudation. Notwithstanding this, changes in lithology have exerted the strongest influence on all rivers at all times." (Weber 1967).

A special case of epigenesis is shown where an older valley is covered by later sedimentation resulting from climatic or tectonic causes to such an extent that with renewed degradation the river can no longer regain its original bed and consequently is cutting a new bed into adjacent fresh and harder bedrock. A well-known example of this situation is presented by the Rhine near Schaffhausen/Switzerland. The river was not able to find its original bed again, which had been filled by fluvioglacial gravel, and thus started at a different level by cutting through a hard escarpment of Jurassic limestones encountered below a thick gravel horizon. Such epigenetic valleys are also met in the Vinschgau of northern Italy. The main causes were large lateral debris cones from tributaries which had frequently deflected the Adige river to the opposite side of the valley where it had to incise itself into solid bedrock (Fischer 1964). Typical for epigenetic valley complexes are also those valleys which start off in a wide shallow valley and in their lower reaches enter rather suddenly into steeply rising bedrock in order to join the equally incised main river in a deep junction gorge. There is frequently a connection between these types of epigenesis and those of the type encountered at the Rhine Cataract near Schaffhausen. Such a set-up is also present in the lowermost reach of the Inn prior to entry into the Danube, i.e., the Vornbach narrows between Schärding and Passau (Louis 1961). In the case of deep valley fills and rejuvenations of valleys or more than 100 m, crustal movements have to be taken into account as additional or even main causes.

Transverse Overflow Valleys. Valleys of this type are frequently developed in glacial regions. They are formed along a ridge such as that of a terminal moraine, behind which initially a melt water lake was in existence. At the lowest point of the ridge the lake water started to flow across. Once the ridge has been cut across completely, the lake, which had already started to become silting up by bedload-laden creeks and rivers, begins to run dry. This is the case of centrifugal discharge in the vicinity of the Alps and the northern inland ice (cf. Chap. 5.2.4.1 and 6.1). Next to gorges, the form of transverse valley occurring here most frequently is that of the "trumpet valley" of Troll. The negative shape of the trumpet valley downstream changes to the positive form of a debris cone.

Regressive Transverse Valleys. Regressive erosion gradually leads to the establishment of a gorge which on a larger scale eventually can cut across a whole mountain range, leading to the capture and deflection of rivers on the other side. It is not always easy to decide whether a particular valley resulted from regressive erosion or antecedence. Regressive erosion is mostly tied to the larger energy of relief of the capturing river.

River Capture. One phenomenon of regressive erosion is known as "the battle for the divide". The river with the lower basis of erosion will, due to its higher descent, have the greater erosive energy and will attempt to enlarge its own catchment area at the expense of that of other rivers. On capturing an adjacent catchment area the divide is suddenly relocated, and entire valleys may be set dry. On further erosive progress by the new river, the old valley is "beheaded". When looking up-valley, it appears suddenly to run out into open space as the deeper valley intervenes. The old valley can frequently be traced beyond the intervening valley up to its original head. The old land form then will be easily reconstructed.

There are many instances of river capture. In Europe the best-known example results from the conflict between Rhine and Danube. One can almost talk of an incursion of the one river, here the Rhine, into the drainage basin of the other, the Danube. The "conflict" can be traced back into the Pliocene. It was eventually decided in favor of the Rhine which, together with its tributaries, and in par-

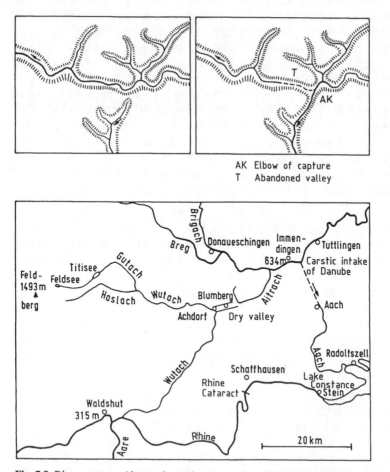

AK Elbow of capture
T Abandoned valley

Fig. 7.8. River capture. *Above* schematic presentation; *below* Wutach/Rhine vs. Danube. (Partly after Wagner 1929)

ticular the Neckar had the higher relief energy on its side. A good example is the capture of the Wutach (Fig. 7.8). The Gutach-Aitrach was the original Pleistocene source stream of the Danube. During the final stages of the Pliocene, the actual sources of the Danube had once been even the Aare and the Alpine Rhine. By regressive erosion, the Wutach, flowing to the Rhine, intruded into the Danube basin and captured the Gutach. This led to the sharp elbow of capture of the Wutach at Achdorf as the Wutach has the deeper erosive basis. At Waldshut it joins the upper Rhine at 315 m a.m.s.l., whereas the Aitrach reaches the Danube at 600 m a.m.s.l. near Immendingen. The present valley floor of the Wutach at Achdorf is 175 m below that of the old dry valley at Blumberg. From Blumberg only a small creek with large gradient now flows in the direction of the Wutach. The regressive erosion in the direction of the residual Aitrach and thus of the Danube is still continuing (Wundt 1953; Wagner 1960).

The regressive erosion of the Moselle during the Pleistocene led to the intrusion into the Meuse basin and the capture of its upper reaches near Toul. This led to an increase in the discharge of the Moselle, which then shaped its further course with large valley meanders under continuous degradation (Kremer 1954).

The rivers on the southern side of the Alps command the greater relief energy due to their communication with the Po plains, of rather low elevation. The Inn, flowing into the Danube, is attacked in the upper Engadin by the source rivers of the Adda. The source spring of the Inn initially did not come from Lake Lunghino, but from the Val Maroz. The Mera has captured the discharge from this valley. From the wide valley of the Inn with its moderate gradient, at the Maloja Pass a steep precipice leads to the Mera in the Bergell.

Erosion controlled by climate and lithology is, however, frequently not the only cause for the relocation of divides. A similar displacement may be achieved also by tectonic processes such as large-scale tilting, which forces the run-off into the direction of the dip, or on a smaller, more localized scale, by doming or subsidence. Subsidence caused by dissolving bedrock, i.e., subsidence of exogene origin, also can lead to river capture, or at least support its development (Weber 1967).

7.2 River History as Part of the History of Civilization

Since historic times the natural formative processes acting on rivers have been modified by human activities. Whether one considers the slow changes of rivers as an indirect consequence of agriculture or the more rapid effects of river regulations, river history from then onward will always form part of the history of civilization. Without further going into intentional modifications of riverside country, a number of facts will be presented here which resulted in response to agricultural technology.

When there is only scarce vegetation cover, any surface water running off will play a considerable role in the transportation of soil matter. Such precipitation will either form innumerable small channels or it will run off in veritable sheet floods. With an increasing gradient, erosion will become stronger. In areas of deforestation and on the so-called "agricultural steppe" the removal of soil extends over vast areas. Even very slight gradients are sufficient here to initiate erosion. Planned or unplanned deforestation campaigns led and still lead to the desertification of large areas. This is presently taking place at an alarming rate in all tropical rain forests.

Deforestation, however, has been going on since man started to cultivate the soil. Since the early Middle Ages in central Europe not only the forests on more hilly country have been cut down, but also those along the flood plains of the rivers, in order to gain new meadows and pastures in the valley. Soil erosion aided by agriculture resulted in an increase of the suspended load of the rivers and has for centuries led to the deposition of thick layers of overbank clays on the flood plains.

In addition to such indirect human influence there have always been direct human modifications of river systems in the form of innumerable mill weirs increasing the deposition of sediment especially along the banks. In the course of time, rivers poor in bedload usually developed meandering courses which retain only a vague hint of the original situation. The annual floods continued to inundate the meadows, but the lack of bedload usually prevented permanent damage. Other than some local protective measures, no necessity for river regulations was seen for a long time. Riversides once established were often maintained for centuries. There was the impression of a natural river and a new state of equilibrium had in fact been established. In contrast to the Alpine rivers and the larger rivers outside this area, which remained untamed right into the last century, all small or medium-sized rivers with low bedload transport have been in a state of transformation for a long time. Large-scale river training carried out at the turn of the century of rivers without bedload transport thus cannot be seen as first attempts to utilize water economically, rather they are only part of a direct line of hydraulic engineering developments since the Middle Ages.

8 Investigations in River Morphology

8.1 Methods

Methods and extent of investigations in river morphology and the measures thus derived for the stabilization of degraded river stretches require a basic understanding of the processes of bed formation, from both a geological as well as a river morphological and engineering point of view (Lane 1955). Increased cooperation between the geologist and engineer has proved to be beneficial, even mandatory in this context.

Next to the geomorphological analysis of the development of a valley and its river system, three different hydrological methods in particular have proved suitable for investigating bed erosion.

8.1.1 Hydrographs of Annual Mean Water Level at the Gages

For a first reconnaissance investigation of an erosive reach, it suffices to follow the changes in the bed level at a few localities. For such a localized investigation, the hydrographs of the annual mean water level at certain official gaging stations (Fig. 8.1) are highly suitable, provided that such data are available on a continuous long-term basis for the river reach under study.

In Bavaria in some areas such records are available since 1826 and actual flow analysis can thus extend over a period of over 160 years, provided that the particular station had not fallen victim to measures of river regulation or that the records have not been discontinued because of cost reasons. The data obtained from analysis of the gage readings provide only preliminary conclusions on the behavior of the river course because they relate only to the particular location of the gage. Taking for example the lower reach of the Iller at Dietenheim (river-km 22.7), it can be observed that there was initially a phase of aggradation following the regulations upstream, until in 1911/12 the excessive width of the passage under the bridge was reduced and river training was continued downstream. In response thereto, degradation set in, which was further increased by a diversion weir ($Q = 80 \text{ m}^3/\text{s}$) constructed in 1928/30. Despite the construction of three sustaining barrages in the years 1958, 1960, and 1969 the erosion could not be stopped. The causes for this situation can be sought firstly in the complete retention of bedload in the upstream string of reservoirs, and secondly in the removal of the alluvial gravel cover of the river bottom in the downstream direction. The Tertiary sublayer, an easily erodible shale/sandstone alternation, was laid bare over

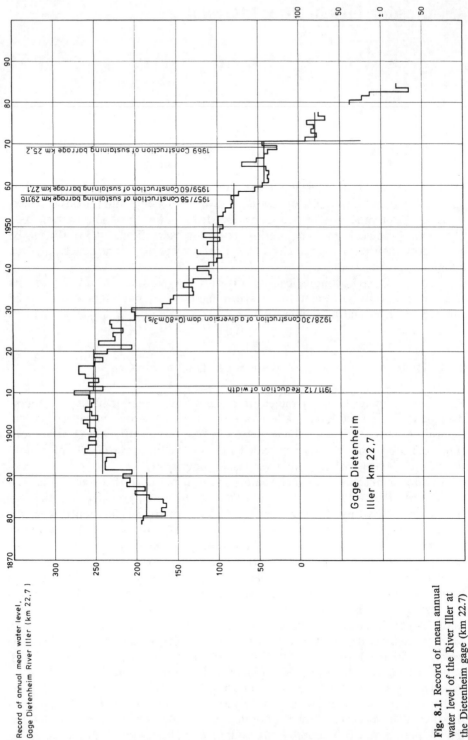

Record of annual mean water level,
Gage Dietenheim River Iller (km 22,7)

Fig. 8.1. Record of mean annual water level of the River Iller at the Dietenheim gage (km 22.7)

a considerable extent, offering little resistance to vertical erosion. The construction of the next sustaining barrage was completed in 1987 and the rehabilitation of the complete remaining stretch cannot be avoided in the long term.

8.1.2 Cross Section Surveys

A more detailed picture of the changes in bed level, and especially their development in the longitudinal section, is provided by cross-sections. In surveyed and calibrated rivers these are carried out regularly at the 200 m markers. For presentation of any changes in elevation along the longitudinal section, the mean bed level or the thalweg may be used. The mean bed level is a calculated value not encountered in nature. It is found from the area defined by the observed bed profile, the two vertical lateral boundaries and an assumed horizontal line. The vertical boundaries usually are fixed by the right and left bench mark of the river cross-section. The mean bed levels, however, are only comparable with each other if the lateral boundaries have not been shifted. The thalweg is the line connecting the

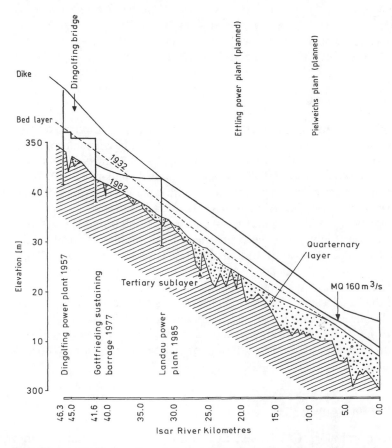

Fig. 8.2. Longitudinal section of the Isar River from Dingolfing to the mouth

deepest points of each cross-section along the course of the river. The cross-sections along the longitudinal section of the lower reach of the Isar are presented in Fig. 8.2. When the thalwegs of a number of cross-section surveys are plotted together with boreholes in the river in a longitudinal section, the excavation and the final penetration of the Quarternary layer, and the subsequent downcutting in the underlying Tertiary sediments can be traced.

For the localization of a break or discontinuity in gradient (knick point) as a base level in an erosive wedge, the more equalized mean bed level is more suitable.

From the cross-section records finally the cumulative mass curve can be calculated and plotted. The current position of the knick point of an erosional stretch is marked in this curve.

8.1.3 Recording of Mean Low Water Level

Periodically repeated recordings of the water level at low discharge are also a suitable method for investigating erosive stretches. In contrast to the recording of the cross-sections only one point has to be surveyed. The work effort and thus the procedural costs are correspondingly lower. Care has to be taken that the fixing of the low water level is undertaken at about the same discharge, so that comparability with the previous measurements is assured (Fig. 8.3).

8.1.4 Methods of Calculation

For the calculation of erosion rates and the maximum degradation in erosive stretches by now numerous models and modes of calculation have been proposed. Amongst these are, e.g., Schoklitsch (1962), Mostafa (1955), Tinney (1962), Komura and Simons (1967), Aksoy (1965), Snishenko and Kopaliani (1984).

In Table 20 the results of computations and field experience of the degradation of the River Nile below the High Aswan Dam are compared.

It shows a considerable discrepancy between the calculated and the actually observed degradation. Only when the properties of the sediments forming the riverbed do not change downward and in the downstream direction, will the calculations yield satisfactory results.

Only few river beds, however, are homogeneous and consequently when there are changes of the layer, especially from Quarternary gravel to fine-grained Tertiary sands and cohesive silts, any further degradation is no longer suitable for calculation and prediction. The only method remaining is then careful observation of the progressive degradation with the aim of instigating the required protective measures at an early stage.

8.1.5 Aerial Photogrammetry

Changes in the morphology of channels as a consequence of human activities such as regulation, reservoir construction, gravel dredging, or diversion of flow

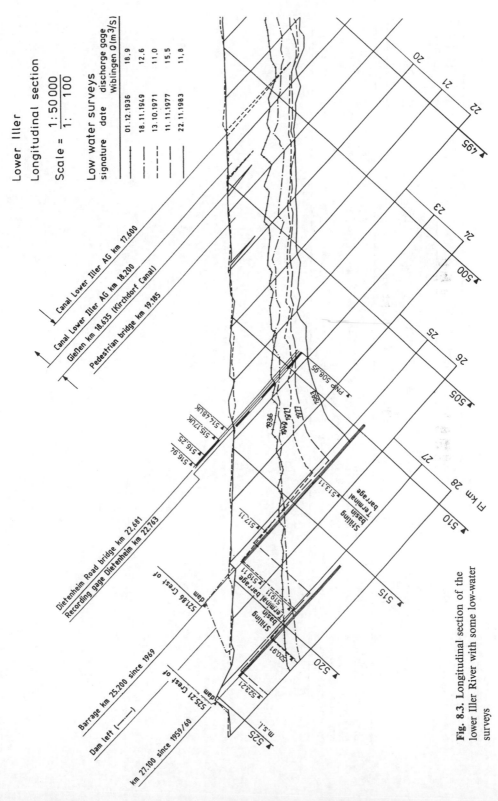

Fig. 8.3. Longitudinal section of the lower Iller River with some low-water surveys

Table 20. Degradation of the River Nile. (After Shalash)

Downstream of	Distance from Low Aswan Dam (km)	Calculated estimates of maximum degradation by						Hydroproject		Field results up to 1982
		Mostafa 1957	VBB 1960	Shalash 1965	Simons 1965	Hydro-project 1973	Shalash 1974	1976	1977	
		Drop in water level (m)								
High Aswan Dam	−6.5	8.50	3.0−4.0	2.0−3.0	−	3.0	1.37	5.3	3.0	0.61
Esna Barrage	167	8.00	3.0−4.0	2.0−3.0	3.50	3.5	1.01	7.0	2.5	0.80
Naga-Hammadi Barrage	360	7.00	3.0−4.0	2.0−3.0	−	3.5	1.37	11.0	4.0	0.99
Asiut Barrage	545	6.50	3.0−4.0	2.0−3.0	−	3.0	2.90	−	8.0	0.73

can be recorded and illustrated by comparison of aerial photos. For this purpose two or more aerial photos taken at greater time intervals, preferentially after major floods, are superimposed and the outlines traced onto a joint map. The changes in the channels may be illustrated by differences in hatching (Fig. 5.8) or by different colors, as in the case of the Brenta (Castiglioni and Pellegrini 1981).

8.2 Examples for Degradation of River Beds Below Barrages and Weirs

The development of degradation below hydraulic structures is influenced mainly by the degree of sediment (bedload) trapping in the reservoir, by the management of discharge in the tailwater and by the natural conditions of the downstream riverbed (layer and sublayer composition and thickness, effects of former training, channel dimensions, possibilities of armoring and barring).

In Table 21 some examples of observed degradations below dams (1 – 2) and hydropower plants (3 – 9) are presented. Nevertheless, local scours can attain much larger depths than the mean values given in the table. Whereas the scour directly below the Saalach dam with a depth of 15 m can be controlled by local measures at the foundation of the structure, the progress of degradation in the following three examples shows the necessity of sustaining barrages.

The 8-m-deep erosional channel of the river Inn below Neuötting developed within only a few years along an unsustained reach between two power plants (Fig. 8.4). The downcutting in the Tertiary sublayer after penetrating the alluvial layer was so heavy, that the rapid construction of an additional supporting power plant became imperative (Fig. 8.5).

The situation developing in the lower Isar near Dingolfing is shown in Figs. 8.6 and 8.7. Two causes have to be named here, the influence of former training and a severe degradation setting in after the construction of the Dingolfing barrage in 1957. This process started about 900 m below a provisional weir and, according to the latest measurement in 1985, has also attained a depth of about 15 m below mean water level.

Below the power plant Großaitingen, on the Wertach, the degradation in the narrow regulated riverbed led to the collapse of the central pillar of a road bridge (Figs. 8.8 and 8.9).

8.3 Possibilities for Rehabilitation of Degradation Stretches

8.3.1 Derivation of Technological Approaches from the "Swiss Formula"

For the general investigation of technological approaches, the bedload transport formula of Meyer-Peter and Müller (1949), the so-called Swiss Formula, will be used as basis (cf. Chap. 4.2.7).

Table 21. Examples of bed degradation in Bavarian rivers

River	Dam, barrage	Year completed	km (from mouth)	Drainage area km²	Mean long-term discharge m³/s	Yearly flood m³/s	Period of observation yr	Degradation Depth m	Degradation Length km	Bed material	Solution
Isar	Sylvenstein	1959	224.4	1138	15.1	125	24	0.90	24	Gravel, sand	4)
Saalach	Reichenhall	1913	20.7	940	38.9	299	47	4.60	18	Sand, gravel	4)
Inn	Jettenbach	1924	128	12250	365	1360	54	2.50	28	Gravel, Tertiary sand	3)
Inn	Neuötting	1951	91.1	13150	370	1270	20	4.50	8 1)	Sand (Tertiary)	2)
Isar	Dingolfing	1957	46.3	8467	167	489	14	2.80	46	Sand, (Tertiary)	2)
Iller	Dietenheim	1969	25.2	1900	55	400	6	1.00	7.6	Sand, (Tertiary)	3)
Wertach	Schwabmünchen	1956	28.3	948	19.7	154	5	1.80	14.5	Sand, (Tertiary)	2)
Danube	Faimingen	1965	2545.5	11315	159	640	12	1.00	16	Gravel, sand	2)
Danube	Ingolstadt	1971	2459.2	20001	308	990	14	1.80	9	Sand, silt (Tertiary)	3)

1) Limited by next weir. 2) Already fixed by a new weir. 3) Weir provided. 4) Only local projects; final solution not yet determined.

Fig. 8.4. Degraded Inn River downstream of the Neuötting power plant (km 91.1). (Photo by J. Schupp)

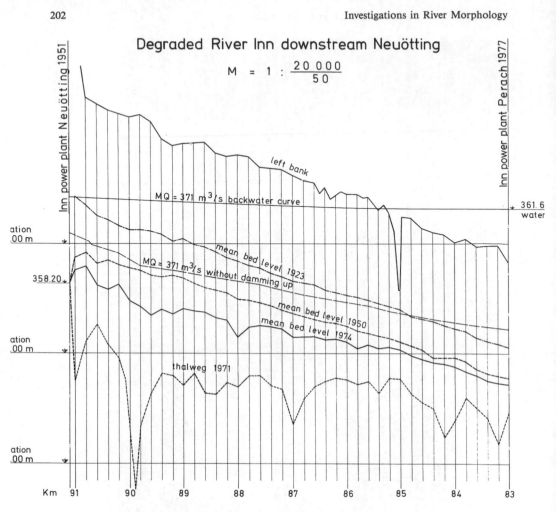

Fig. 8.5. Longitudinal section of the River Inn downstream of Neuötting

All three terms of this equation represent shear stresses. If the simplified Eq. (67) is converted to

$$\alpha \cdot h \cdot J = \beta d_m + \varepsilon g_s''^{2/3} \, , \tag{67a}$$

then the first term represents the active shear stress, the second the boundary or critical shear stress at which movement of bedload and thus destruction of the bed layer will start, and the third the shear stress available for bedload discharge.

When the supply of bedload is cut off, as for instance below barrages, the third term yields zero. This means that the share of kinetic energy available for bedload transport can now no longer be consumed by bedload, and will therefore be fully taken over by the bed layer, from thus leading to a gradual dissappearance of the Quarternary gravel cover and penetration of this layer. The underlying, usually fine-grained Tertiary sediments offer little resistance to the progressing degradation. In order to re-establish the previous equilibrium of the bed level, dif-

Fig. 8.6. Degraded River Isar below Dingolfing bridge (km 45.5). (Photo by W. Binder)

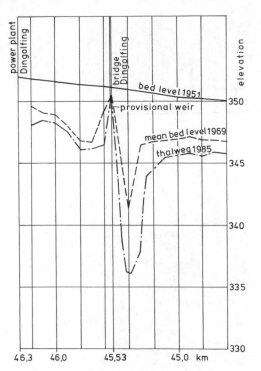

Fig. 8.7. Longitudinal section of the Isar with thalweg

ferent solutions are feasible. If we consider the simplified bedload Eq. (67 a), there is the possibility of reducing on the left side the slope J at unchanged grain diameters.

On the right side the grain size d_m may be increased without changing the actual gradient. In this way, the actual resistance of the bed layer can be increased.

8.3.2 Rehabilitation by Increasing the Drag

For bolstering the resistance of the layer there are three possibilities. Firstly the stabilization of the layer by armoring, secondly the construction of sills or a combination of these methods, and thirdly the addition of bed material.

8.3.2.1 Stabilization by Armoring

In engineering terms armoring implies the application of a sufficiently dimensioned artificial cover of stone material in order to protect the bed layer from scour. The necessary grain size can be determined by the diagram of Shields (Fig. 4.13). The armored layer must be underpinned by a filter-type sublayer in order to prevent the extraction of the fine-grained Tertiary material. Huppmann (1976)

Fig. 8.8. Collapsed bridge downstream of the Grossaitingen power plant (Wertach River km 17.4); photo by Lech-Elektrizitätswerke AG, Augsburg

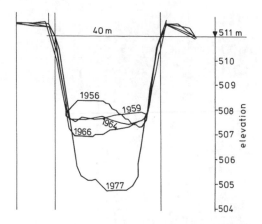

Fig. 8.9. Development of downcutting 400 m downstream of the bridge (Wertach km 17.0)

investigated whether the "classical" gravel filter layer could be replaced by modern fiber filter mats. He found that neither mats of artificial fibers nor of felt alone would be sufficient. A combination of the two materials, however, appears to be suitable, i.e., a felt matting on a coarser carrying fabric, the fabric taking up the mechanical stresses and the felt accounting for the filter action. It must be stressed that on installation these mats have to be weighted down, for instance by concrete blocks, or secured by steel hooks or anchors. The armored layer must have a minimum thickness of 1.2−1.5 m, allowing for the fact that during floods up to 1 m of the bed layer may be mobilized. The installation has to be carried out in two layers under water. Great care and uniform distribution have to be ensured, as at any weak points the mats will easily be rolled up by floods. Checking during installation is difficult. The risks inherent in this system, which is technically feasible and of little environmental impact, must not be underrated. It has been applied in hydraulic engineering for some time, but always only locally on damaged stream banks, in scours, on bridge pillars, and below weirs. So far, no experience with the application over longer stretches has been reported. Special care has to be taken where fine-grained layers such as the Tertiary Flinz (silts and sands) have already been uncovered.

8.3.2.2 Bottom Belts and Sills

A support and an additional protection for the artificial carpet described above might be feasible by cross-structures installed at larger intervals. These are the so-called sills or ridges in the form of "rock sausages", concrete ridges, or concrete-filled fiber hoses. According to Hartung (1973), it must be taken into account that possibly there will still be some movement of these belts, leading to shorter spacing. In Switzerland this type of belt is known as "cross-ridges". The ridges will lead to ripples and dunes as encountered in a natural river. However, as they are larger, they will result in stronger turbulences and consequently greater erosive effects, threatening the stability of the river bed between the belts. Without the installation of a continuous protective cover layer between such belts, this method will not be applicable to finer-grained bed material. Mayrhofer in particular investigated the technology of bottom belts. Based on model considerations (Ziller) and practical tests in the Inn and Raab rivers, he generally advised against the construction of bottom belts. This is mainly because of the only partial effectiveness of these structures at the discharge to be expected especially during floods.

8.3.2.3 Addition of Bed Material

The artificial feeding of bed material as a method of engineering is currently practiced on the upper Rhine below the barrages Gambsheim and Iffezheim (Felkel 1970, 1980). Annually about 167000 m^3 (1978/85) of gravel with a mean particle diameter of 22 mm are fed to the river (Felkel 1987). This has not led to undesired sedimentation in the downstream section until now but it is envisaged that eventually the material will have to be dredged out again in order to ensure the navigability of the river. Here the problem was not an imminent or already existing penetration of the bottom cover but the re-establishment of the original level of the degraded river bed below the last barrage. Before using this method somewhere else, it first has to be examined at all whether the conditions encountered in the upper Rhine are applicable to the section to be rehabilitated. An important difference could lie in the properties of the bed layer. Whereas the upper Rhine is cutting into an over 120 m thick alluvial sequence of gravels and sand, in most of the degraded Bavarian rivers the Quarternary layer is almost removed and the easily erodible Tertiary sublayer has already been attacked. A direct feeding of bedload onto this layer without preventive measures on the bottom itself in the form of filter mats, for example, could actually have a negative effect, due to the rubbing and cutting action exerted by the bedload rolling and saltating along the bottom. Whereas in the case of the Rhine the bedload transport inhibited by the barrages had to be re-established over a sufficiently thick gravel layer, in the Bavarian rivers such a heavy gravel layer would have had to be built up most painstakingly and under difficult conditions over the uncovered Tertiary layer. The gravel layer would have to be 1.2 – 1.5 m thick so as to offer an adequate protective influence. Whereas along the Rhine the supply of gravel would not be overly expensive, in other valleys used by agriculture, industry, and traffic lanes, this will not be the case. The gravel beds here are comparatively thin and the areal requirements would be larger than along the Rhine, where gravel thicknesses of 20 m are by no means unusual (Felkel 1970). Dredging has gone down to 40 – 50 m in recent years.

In the case of the Rhine, addition of bed material by means of bottom-release barges was found to be the most practicable method whereas the low water levels in smaller rivers preclude the use of barges. Special distributing devices would have to be developed first.

It must also be kept in mind that contrary to the situation in the Rhine the added bedload material, minus the attrition losses, might lead to undesirable deposits above the mouth of the next river due to the decrease in gradient. The ensuing gravel dredging at the mouth and its re-addition at the original point would prove costly and produce a considerable disturbance of the valley by new roadways and related environmental harassment by noise and dust.

It should be noted that bedload transport will set in only at higher discharge and that during floods, within a few days, it may exceed even the mean annual transport rate. According to the present transport capacity, a potential need for new gravel placements of tens of thousands of tonnes must be envisaged.

Felkel (1970) reported that in the case of the upper Rhine the feeding of bed material results in lower costs than the alternative construction of barrages or of armoring the river bed. This would not be the case with the Bavarian rivers, where an addition of bed material can be recommended only if sufficient gravel of suitable grain size composition is available within a short distance, and where this would not lead to undesired deposition downstream. Consequently, this solution is presently practised only in the case of the Saalach River below the Bad Reichenhall dam, whence it is intended to transport the material to the degraded Salzach.

8.3.3   Rehabilitation by Reduction of Flow Velocity or Gradient

According to Eq. (67 a) a reduction of the active shear stress can result from the reduction of the gradient or of the flow velocity.

8.3.3.1   Widening of the Channel

Bringing down the flow velocity to the permissible critical velocity calls for either widening the existing channel profile or damming up the river. The critical velocity is found from the grain size distribution of the material covering the river bed. However, in most cases a notable widening of the cross-section in order to remain below the critical velocity is virtually impossible with the present extent of land use. It would detract from the advantages gained by earlier regulative measures.

Other important arguments speak against a widening of the bed profile, as pointed out by Mayrhofer (1970). Each river shows large fluctuations in discharge, and the profile would have to be designed for the greatest flood run off anticipated. During low or mean water levels the bottom then theoretically would have a very shallow water cover over the whole width. In reality, the profile, however, would soon be occupied by deeper individual channels which would be relocated during every flood. Because of the insufficient supply of bedload, no braided system would develop as in natural stretches of redeposition, and after even a short time the discharge would be concentrated along a stretched channel, further degrading it. The adjacent areas would soon be covered by bushes which would have to be cleared continuously in order to maintain the discharge capacity of the channel. This situation and the large space demand preclude profile widening, at least on its own, from being a practicable solution.

8.3.3.2 Drops and Weirs with Low Heads

A reduction of the gradient and with it of the active shear stress could be achieved by means of drops. In natural rivers with their low gradients compared to torrents, drops would be out of place as at flood levels they are not sufficiently effective from the hydraulic point of view.

If it is not possible to reduce any excess of energy by enforcing a "hydraulic jump", the resulting undulating discharge, the excessive velocities and macro-turbulences will expose the downstream river bed to heavy stress and thus dangerous erosion (Hartung 1973).

When planning drops, great care has to be taken to obtain the "minimum drop" required for hydraulic efficiency and thus a safe energy dissipation will be achieved.

8.3.3.3 Stone Ramps

Slope reduction can also be considered by initiating natural rapids by rough rock ramps, the so-called Schauberger ramps or bottom ramps. From the hydraulic point of view they also represent drop structures. When very shallow (1 : 15) they may be applied into a specific discharge of $9 \, \mathrm{m}^3 \, \mathrm{s}^{-1} \, \mathrm{m}^{-1}$ as elements of local protection of the layer in small- to medium-sized rivers (Knauss 1977). However, the acceptable degree of specific discharge will be by far exceeded in larger rivers and the hydraulic efficiency of the ramps will no longer be guaranteed at higher discharge. They are thus only useful within the above limits: the hydraulic engineer has come to the end of the "natural cures" available to him (Hartung).

8.3.3.4 Sustaining Barrages and Power Plants

In order to achieve the essential drop for effective energy dissipation, the crest of the drop in the headwater must be lifted, and the river must be dammed. The sustaining barrage achieves an increasing of the profile area and a decreasing of slope and velocity in the headwater; the discharge in the tailwater is adapted to the erodible layer with regard to velocity, turbulence, and forming of waves. This barrage is a fixed weir, with the disadvantage that the overflow level is not variable, but depending on the momentary discharge. In order to control even high floods without inundation, high dykes would be required. The steadily changing water level would inhibit an ecologically satisfactory landscaping in the reservoir.

The significant disadvantage of the fixed sustaining barrages are the high embankments, and the continuous fluctuations of the head water and the level of the lake. These may be avoided by proceeding to the further expense of incorporating sluice gates in the barrages to control the discharge so that water levels may be kept fairly constant.

8.3.3.5   Low-Head Hydro-Power Plants

Such structures are, as it were, movable weirs combined with power stations, damming up far across the foreland. The backwater level is also ensured during floods by lowering of the gates. Whereas in the case of sustaining barrages an economic generation of power will rarely be possible due to the low permanent water level this is considerably improved in the low-head power stations due to the increased permanent level of the backwater.

8.3.4   Conclusions

The critical evaluation of eight various solutions has shown that in larger rivers virtually only the two last ones, i.e., the sustaining barrage schemes and the low-head hydro-power plants, are worthy alternatives for a comprehensive concept. The improvement of excessively low bed resistance over longer reaches by armoring or bottom belts promise no expectation of long life. The addition of gravel is also eliminated because of the already locally exposed layer of easily erodible sediments, the extensive and environmentally unacceptable transport organization high costs, the logistics of distribution the supply, and the risk of uncontrolled deposition in downstream sections. Because channel widening, on the other hand, would establish a stretched, actively eroding channel due to the considerable fluctuations of discharge and the lack of bed material supply and would cause a new set of morphological problems, only the solution by drops remains.

In developing such control works the use of minor structures should be avoided because low weirs are not sufficiently effective from the hydraulic point of view. The better strategy for permanent protection of the river bed from further degradation would seem to be sustaining barrages with good hydraulic conditions or, if we include energy-economic considerations, the construction of a series of power plants.

References

Abrahams AD (1972) Drainage densities and sediment yields in eastern Australia. Aust Geogr Stud 10:19–41

Abrahams AD (1984) Channel networks: a geomorphological perspective. Water Resour Res 20:161–188

Abrahams AD (1984/85) Spatial dependency of hydraulic geometry exponents in a subalpine stream – a comment. J Hydrol 75:389–393

Abrahams AD, Ponczynski JJ (1984/85) Drainage density in relation to precipitation intensity in the USA. J Hydrol 75:383–388

Ackers P (1988) Alluvial channel hydraulics. J Hydrol 100(1/3):177–204

Ackers P, Charlton FG (1970) Meanders geometry arising from varying flows. J Hydrol 11:230–252

Ackers P, Charlton FG (1975) Theories and relationships of river channel patterns. J Hydrol 26(3–4):359–362

Ajbulatov N, Boldyrev V, Griesseier H (1961) Das Studium der Sedimentbewegung in Flüssen und Meeren mit Hilfe von lumineszierenden Farbstoffen und radioaktiven Isotopen. Peterm Geogr Mitt 3. Quartalsh: 177–186, 4. Quartalsh: 254–263

Aksoy S (1965) An analytical study of the river-bed degradation downstream of large dams. IAHR XIth Congr Leningrad, 10 pp

Alfieri N, Arias PE (1958) Spina. Die neuentdeckte Etruskerstadt und die griechischen Vasen ihrer Gräber. Hirmer, München, 66 pp

Allen JRL (1965) A review of the origin and characteristics of Recent alluvial sediments. Sedimentology, Amst, 5(2):89–191

Allen JRL (1979) Studies in fluviatile sedimentation: an elementary geometrical model for the connectedness of avulsion-related channel sand bodies. Sediment Geol 24(3/4):253–267

Anastasi G (1984) Geschiebeanalysen im Felde unter Berücksichtigung von Grobkomponenten. Mitt VA Wasserbau, Hydrologie u. Glaziologie, ETH Zürich 70:999

Andreasen AHM (1958) Bestimmung der Korngröße durch Siebung und Sedimentation. Ber Dtsch Keram Ges 35:150–153

Arafa FAK (1978) Sohlausbildung und Gesamtsedimenttransport in natürlichen Gewässern unter Berücksichtigung der Kornverteilung und der Dichte des Sohlmaterials. Mitt Inst Wasserbau Wasserwirtschaft TH Aachen 23:170 pp, A 48 pp

Arnborg L (1957a) Erosion forms and processes on the bottom of the river Angermanälven. Geogr Ann 39:32–47

Arnborg L (1957b) The lower part of the river Angermanälven. Geogr Inst Univ Uppsala 1:181–247

Aulitzky H (1984) Vorläufige, zweigeteilte Wildbachklassifikation. Wildbach- Lawinenverbau 48(SH):7–60

Bagnold RA (1960) Some aspects of the shape of river meanders. US Geol Surv Prof Pap 282:135–144

Bagnold RA (1966) An approach to the sediment transport problem from general physics. US Geol Surv Prof Pap 422-1:37 pp

Bagnold RA (1973) The nature of saltation and of "bedload" transport in water. Proc R Soc Lond, Ser A, 332:473–504

Baker VR (1987) Paleoflood hydrology and extraordinary flood events. J Hydrol 96:79–99

Band LE (1986) Topographic partition of watersheds with digital elevation models. Water Resour Res 22(2):15–24

Bardsley WE (1981) Note on two limit distributions of bedload movement. J Hydrol 52(1–2): 165–169

Barnes HH (1967) Roughness characteristics of natural channels. US Geol Surv Water Supply Pap 1849:213 pp

Bartz J (1961) Die Entwicklung des Flußnetzes in Südwestdeutschland. Jh Geol Landesamt Baden-Württemberg, Freiburg/Br, 4:127–135

Batel W (1971) Einführung in die Korngrößenmeßtechnik. Springer, Berlin Heidelberg New York, 3. Aufl, 214 pp

Bauer F (1960) Entwicklung des Entnahmegerätes für Schwebstoffmessungen. Comm d'Erosion Continentale de l'AIHS 53:23–25

Bauer F (1965) Der Geschiebehaushalt der bayerischen Donau im Wandel wasserbaulicher Maßnahmen. Die Wasserwirtschaft, Stuttgart, 55(4):106–112; 5:145–154

Bauer F (1968) Die Verlandung in natürlichen Seen, Talsperren und Flußkraftwerkstreppen. Festschr Kongr Wasser, Berlin, 12 pp

Bauer F, Burz J (1968) Der Einfluß der Feststofführung alpiner Gewässer auf die Stauraumverlandung und Flußbetteintiefung. Die Wasserwirtschaft, Stuttgart, 58(4):114–121

Bauer L (1952) Hydrologie des Flußgebietes von Unstrut und Gera unter besonderer Berücksichtigung der Hochwassererscheinungen und des Einflusses von Kahlschlägen auf die Wasserführung. Diss Univ Jena, Jena, 76 pp

Bauer L (1961) Zur Hydrogeographie des Schwarza- und Rodagebietes – ein Beitrag zur Gewässerkunde und Gewässerpflege in Thüringen. Arch Naturschutz Landschaftsforsch, Berlin (Ost), 1(2):99–141

Bauer L, Tille W (1966) Die Sinkstofführung der Fließgewässer des Unstrut-Gebietes (Ihre hydrogeographische und landeskulturelle Problematik). Pet Geogr Mitt, Gotha 2:97–110

Bauer L, Tille W (1968) Über die hydrogeographische Differenzierung des Sinkstofftransportes thüringischer Fließgewässer und die Beziehungen zur Bodenerosion. Peterm Geogr Mitt, Gotha/Leipzig 112(1):37–42

Baulig H (1948) Le problème des méandres. Bull Soc Belge Étud Geogr 17:103–143

Bayerische Landesstelle für Gewässerkunde, München (1972a) Statistische Auswertung von Schwebstoffmessungen. Abschlußbericht zu einem Forschungsvorhaben, gefördert von der DFG München, 34 pp

Bayerische Landesstelle für Gewässerkunde (1972b) Die Schwebstofführung bayerischer Flüsse. Eigenverlag: amtliche Publ, München, 21 Tab, 40 Beil

Becchi I, Billi P, Tacconi P (1981) Analysis of a simple suspended load integrating sampler. Proc Florence Symp IAHS-AISH 133:527 pp

Bechteler W (1980) Stochastische Modelle zur Simulation des Transportes suspendierter Feststoffe. Wasserwirtschaft 70(5):191–196

Bechteler W (1985) Modelle zur Ermittlung der Schwebstoffkonzentrationsverteilung in Fließgewässern. Hydraulik Gewässerkde 44:47–78

Beckinsale RP (1969) River regimes. In: Chorley RJ (ed) Water, earth and man. Methuen, Lond, pp 455–471

Behrmann W (1913) Die Oberflächengestaltung des Harzes. Forsch Dtsch Landes-Volksk 20:193

Bensing W (1966) Gewässerkundliche Probleme beim Ausbau des Oberrheins. DGM, Koblenz, 10(4):85–102

Behrmann W (1933) Morphologie der Erdoberfläche. In: Klute F (ed) Handbuch der geographischen Wissenschaft. 1. Teil: Physikalische Geographie. Akademische Verlagsgesellschaft Athenaion mbH, Potsdam, pp 356–537

Beven KJ, Wood EF, Sivapalan M (1988) On hydrological heterogeneity – catchment morphology and catchment response. J Hydrol 100(1/3):353–375

Bikshamaiah G, Subramanian V (1980) Chemical and sediment mass transfer in the Godavari River basin in India. J Hydrol 46(3–4):331–342

Biksham G, Subramanian V (1988) Sediment transport of the Godavari River Basin and its controlling factors. J Hydrol 101(1/3):275–290

Birot P (1961) Réflexions sur le profil d'équilibre des cours d'eaux. Z Geomorph NF, Berlin, 5, H 1:1–23

Blache J (1939/1940) Le problème des méandres encaissés et les rivières lorraines. J Geomorph, NY, 2:201–212; 3:311–331

Blau E (1957) Die Bedeutung und die Aufgaben der Geschiebeforschung für den Flußbau. WWT 7(10):401–407

Blau E, Krause H (1967) Ein neuer Geschiebefänger mit Saugdüse, Einspül- und Wägevorrichtung (DWP). WWT 17(5):171–174

Bluck BJ (1974) Structure and directional properties of some valley sandur deposits in southern Iceland. Sedimentology, Oxford, 21:533–554

Blyth K, Rodda J (1973) A stream length study. Water Resour Res 9:1454–1461

Bogárdi J (1956) Über die Zu- und Abnahme des Schwebstoffgehaltes in den Flüssen mit der Änderung des Abflusses. Die Wasserwirtschaft, Stuttgart, 47(3):59–66

Bogárdi J (1959) Neuere Erkenntnisse auf dem Gebiet der Geschiebeforschung. ÖWW, Wien, 11(12):286–293

Bogárdi J (1959) Neue Parameter und Invarianten bei der Bestimmung der Geschiebeförderfähigkeit. Die Wasserwirtschaft, Stuttgart, 49(12):314–320

Bogárdi J (1968) Bestimmung der Grenzzustände bei der Geschiebebewegung. Die Wasserwirtschaft, Stuttgart, 58(7):205–212

Bogárdi J (1974a) Feststoffprobleme, Theorie und Praxis. ÖWW, Wien, 26(7/8):153–163

Bogárdi J (1974b) Sediment transport in alluvial streams. Akadémiai Kiadó, Budapest, 826 pp

Bogomolov GW (1958) Grundlagen der Hydrogeologie. D Verlag d Wissensch, Berlin, 178 pp

Böhmer G (1922) Die Flußdichte im Gebiete der mecklenburgischen Seenplatte und ihrer Vorländer. Diss (Kurzf) EM Arndt Univ, Rostock, 16 pp

Bonnefille R (1963) Essais de synthèse des lois de début d'entraînement des sédiments sous l'action d'un courant en régime continu. Bull CREC, Chatou, 5:67–72

Brehm J, Meijering MPD (1982) Fließgewässerkunde. Biol Arb Bücher Bd 36, Quelle und Meyer, Heidelberg, 312 pp

Bremer H (1959) Flußerosion an der oberen Weser. Göttinger Geogr Abh, Göttingen, 22:192 pp

Bremer H (1960) Neuere flußmorphologische Forschungen in Deutschland und ausgewählte Probleme der Flußmorphologie deutscher Ströme. Ber Dtsch Landeskde, Bad Godesberg, 25(2):283–299

Bremer H (1968) Der Fluß als ein Gestalter der Landschaft. Geogr Rdsch, 20. Jg, H 10:372–381

Bremer H (1971) Flüsse, Flächen- und Stufenbildung in den feuchten Tropen. Würzburger Geogr Arb 35:194 pp

Bremer H (1972) Flußarbeit, Flächen- und Stufenbildung in den feuchten Tropen. Z Geomorph, Suppl-Bd 14:21–38

Bremer H (ed) (1985) Fluvial geomorphology. Z Geomorphologie (Suppl) 55:150 pp

Brenken G (1959) Versuch einer Klassifikation der Flüsse und Ströme der Erde nach wasserwirtschaftlichen Gesichtspunkten. Diss TH Karlsruhe, 42 pp

Bretschneider H (1968) Einige Ähnlichkeitsbetrachtungen zur Geschiebetriebformel von Meyer-Peter und Müller. Die Wasserwirtschaft, Stuttgart, 58(9):269–271

Brice JC (1974) Evolution of meander loops. Geol Soc Am Bull 85:581–586

Brinkmann R (1984) Abriß der Geologie. I. Allgemeine Geologie, neu bearb v Zeil W, Enke, Stuttgart, 13. Aufl, 276 pp

Brotherton DI (1979) On the origin and characteristics of river channel patterns. J Hydrol 44:211–230

Browzin BS (1964) Seasonal variations of flow and classification of rivers in the Great Lakes – St. Lawrence Basin. Univ Michigan, Great Lakes Res Div, East Lansing, 11:179–204

Bruk S (1969) Schwebstofführung feinsandiger Wasserläufe. VA Wasserbau Kulturtechnik Univ Karlsruhe, pp 127–156

Bruk S (1973) Variation of river plan forms (braiding, meandering, alluvial fans). Int Semin Hydraul Alluvial Streams, New Delhi, Central Board of Irrigation and Power, vol 1, lecture no 8

Bruk S, Miloradov V (1967) Untersuchung der hydraulischen Zusammenhänge beim Schwebstoff- und Feinsandtransport in der jugoslawischen Donaustrecke. DGM, Sonderh 1967, Koblenz, pp 38–51

Bunte K, Custer SG, Ergenzinger P, Spieker R (1987) Messung des Grobgeschiebetransportes mit der Magnettracertechnik. DGM 31, H 2/3:60–67

Bunza G, Karl J, Mangelsdorf J, Simmersbach P (1976) Geologisch-morphologische Grundlagen der Wildbachkunde. Schriftenr Bayer Landesst Gewkde, München, 11:128 pp

Burz J (1956) Abrieb und Sohlengefälle. DGM, Koblenz, 9(2):29–32

Burz J (1958) Abgrenzung der Schwebstoff- und Sohlenfracht. Die Wasserwirtschaft, Stuttgart, 48(14):387–389

Burz J (1967) Verteilung der Schwebstoffe in offenen Gerinnen. Publ IAHS No 75:279–296

Burz J (1971) Erfahrungen mit der fotometrischen Trübungsmessung. Bes Mitt Dtsch Gewk Jb 35, Koblenz, pp 355–364

Carlston CW (1963) Drainage density and streamflow. US Geol Survey Prof Pap 422C:8 pp

Carlston CW (1965) The relation of free meander geometry to stream discharge and its geomorphic implications. Am J Sci 263:864–885

Carlston CW (1966) The effect of climate on drainage density and stream flow. Int Assoc Sci Hydrol Bull 11:62–69

Carson MA (1984) The meandering – braided river threshold: a reappraisal. J Hydrol 73:315–334

Castiglioni GB, Pellegrini GB (1981) Two maps of the dynamics of a river bed. Erosion Sediment Transport, IAHS Symp, Florence, 3 pp

Cati L (1981) Idrografia e Idrologia del Po. Ufficio Idrografico del Po, Rom, 19:310 pp

Chang HH (1979) Minimum stream power and river channel patterns. J Hydrol 41(3–4):303–327

Chang HH (1984/85) Meandering of underfit streams. J Hydrol 75:311–322

Chang HH (1987) Comment on "modeling alluvial channels" by David R Dawdy and Vito A Vanoni. Water Resour Res 23(11):2153–2155

Chang HH (1988) Fluvial processes in river engineering. John Wiley & Sons, New York, 432 pp

Chang TP, Toebes GH (1971) Geometric parameters for alluvial rivers related to regional geology. Proc 14th Congr IAHR, Paris, 3(C24):193–201

Charlton FG, Benson RW (1966) Effect of discharge and sediment charge on meandering of small streams in alluvium. Symp CWPRS, Poona, 2:285–290

Chien N (1961) The braided stream of the lower Yellow River. Sci Sinica 10(6):734–754

Chitale SV (1970) River channel patterns. Proc Hydrol Div, J Am Soc Civ Eng 96(1):201–222

Chitale SV (1973) Theories and relationships of river channel patterns. J Hydrol 19:285–308

Chorley RJ (1969) Water, earth and man. Methuen, Lond, 588 pp

Chorley RJ (1969) The drainage basin as the fundamental geomorphic unit. In: Chorley RJ (ed) Water, earth and man. Methuen, Lond, pp 77–99

Chorley RJ, Morgan MA (1962) Comparison of morphometric features, Unaka Mountains, Tennessee and North Carolina, and Dartmoor, England. Geol Soc Am Bull 73:17–34

Chorley RJ, Malm DEG, Pogorzelski HA (1957) A new standard for estimating drainage basin shape. Am J Sci 255:138–141

Chow VT (1959) Open channel hydraulics. McGraw-Hill, NY, 680 pp

Christiansen H (1974) Über den Transport suspendierter Feststoffe in Ästuarien am Beispiel der Elbemündung bei Neuwerk. Hamburger Küstenforschung H 28

Church MA (1972) Baffin Island sanders: a study of Arctic fluvial processes. Canada Geol Survey Bull, 216 pp

Colby BR (1963) Fluvial sediments – a summary of source, transportation, deposition and measurement of sediment discharge. US Geol Surv Bull 1181:1–47

Cole WS (1930) Interpretation of intrenched meanders. J Geol 38:423–436

Coleman JM (1969) Brahmaputra River: channel processes and sedimentation. Sediment Geol, Amst, 3:129–239

Collinson JD, Lewin J (1983) Modern and ancient fluvial systems. Spec Publ Int Assoc Sedimentologists 6:584 pp

Corbel J (1959) Vitesse de l'érosion. Z Geomorph, NF, 3:1–28

Cotton CA (1964) The control of drainage density. NZ J Geol Geophys 7:348–352

Cox JC, Harisson SS (1979) Fracture-trace influenced stream orientation in glacial drift, northwestern Pennsylvania. Can J Earth Sci 16(7):1511–1514

Crickmore MJ (1967) Measurement of sand transport in rivers with special reference to tracer methods. Sedimentology 8(3):175–228

Culling WEH (1960) Analytical theory of erosion. J Geol 68(3):336–344

Curry RR (1972) Rivers – a geomorphic and chemical overview. In: Oglesby, Carlson, McCann (eds): River ecology and man. Academic Press, NY Lond, pp 9–32

Czajka W (1958) Schwemmfächerbildung und Schwemmfächerformen. Mitt Geogr Ges, Wien, 100:18–36

Daly RA (1936) Origin of submarine "canyons". Am J Sci, Ser 5, 31:400–420

Daniel JRK (1981) Drainage density as an index of climatic geomorphology. J Hydrol 50(1–3): 147–154

Dantscher K (1942) Die Flüsse und die Erdrotation. Wasserkraft Wasserwirtschaft, München, 37(12):269–274

Da Vinci L (reprint 1828) Del moto e misura dell'acqua. a spese di F Cardinali, Cardinali, Bologna, pp 273—450

Davis WM (1890) The rivers of north New Jersey with notes on the classification of rivers in general. Natl Geogr Mag 2:81—110

Davis WM (1893) The topographic maps of the United States Geological Survey. Science 21:226—227

Davis WM (1903) The development of river meanders. Geol Mag 10:145—148

Davis WM (1909) The Seine, the Meuse, and the Moselle. Geogr Essays, NY, pp 587 ff

Davis WM (1923) River meanders and meandering valleys. Geogr Rev pp 629—630; 1924: pp 504

Dawdy DR, Vanoni VA (1986) Modeling alluvial channels. Water Resour Res 22(9):71—81

de Haar U, Keller R (1979) Hydrologischer Atlas der Bundesrepublik Deutschland. Im Auftr Dtsch Forschungsgemeinschaft, 365 pp

Deutscher Verband für Wasserwirtschaft und Kulturbau (1986) Schwebstoffmessungen. DVWK Regeln Wasserwirtschaft 125:46 pp

De Vries M (1961) Computations on grain sorting in rivers and river models. Delft Hydraul Lab Publ 26

De Vries M (1975) A morphological time scale for rivers. Proc 16th Congr IAHR, São Paulo, 2(B3):17—23

Dietrich B (1911) Entstehung und Umbildung der Flußterrassen. Geol Rundsch, pp 445—454

Dietrich G (1959) Zur Topographie und Morphologie des Meeresbodens im nördlichen Nordatlantischen Ozean. Dtsch Hydrogr Z Erg 3

Dietrich WE, Smith JD (1983) Influence of the Point Bar on flow through curved channels. Water Resour Res 19(5):1173—1192

Dietrich WE, Smith JD (1984) Bed load in a river meander. Water Resour Res 20(10):1355—1380

Diez T (1968) Die würm- und postwürmglazialen Terrassen des Lech und ihre Bodenbildungen. Eiszeitalter Gegenwart 19:102—128

Dillo HG (1960) Sandwanderung in Tideflüssen. Mitt Franzius-Inst TH Hannover 17:135—253

DIN 4049 Teil 1 (1979) Hydrologie Begriffe, quantitativ. Institut für Normierung, Beuth-Verlag, Berl, Neufassung v 1954, 54 pp

Dingman SL (1978) Drainage density and streamflow: a closer look. Water Resour Res 14(6): 1183—1187

Dingman SL (1984) Fluvial hydrology. Freeman NY, 383 pp

Dixey F (1974) Drainage basin form and process; a geomorphological approach, by Gregory KJ and Walling DE. J Hydrol 23(3—4):357—359

Dixey F (1975) Fluvial processes in instrumented watersheds; studies of small watersheds in the British Isles, by Gregory KJ and Walling DE. J Hydrol 25 (1—2):173—175

Dixey F (1975) The work of rivers; a critical study of the central aspects of geomorphology (book review). J Hydrol 27(1—2):169—170

Dornbusch W (1965) Ein neues Gerät zur Bestimmung der Geschiebeführung. Wasserwirtschaft/ Wassertechnik H 9:305—307

Douglas I (1974) The impact of urbanisation on river systems. Proc Int Geogr Union Reg Conf, NZ, Geogr Soc, pp 307—317

Drexler O (1979) Einfluß von Petrographie und Tektonik auf die Gestaltung des Talnetzes im oberen Rißbachgebiet (Karwendelgebirge, Tirol). Münchner Geogr Abh, München, 23:124 pp

Du Boys P (1879) Le Rhône et les rivièrs a lit affonillable. Ann Ponts Chaussées, Ser 5, 18:141—195

Du Buat P (1816) Principes d'hydraulique. Paris

Dudziak J (1981) Die Abhängigkeit der kritischen Schleppkraft von den Transportbedingungen. Wasserwirtschaft 71(7/8):229—231

Düll F (1930) Das Gesetz des Geschiebeabriebes. Mitt Geb Wasserbaus Baugrundforsch, Berlin, 1:62 pp

Dunne T, Leopold LB (1978) Water in environmental planning. Freeman, Reading, 818 pp

Dury GH (1953) The shrinkage of the Warwickshire Itchen. Coventry Nat Hist Sci Soc Prov 2:208—214

Dury GH (1954) Contribution to a general theory of meandering valleys. Am J Sci 252:193—224

Dury GH (1964) Principles of underfit streams. US Geol Survey Prof Pap 452A:67 pp

Dury GH (1970) Rivers and river terraces. Macmillan, NY Washington, 283 pp

Dyck S (1980) Angewandte Hydrologie. I. Berechnung und Regelung des Durchflusses der Flüsse. 2. Aufl, 528 S. II. Der Wasserhaushalt der Flußgebiete. Verlag für Bauwesen, Berlin, 2. Aufl, 552 pp

Dyck S, Peschke G (1983) Grundlagen der Hydrologie. Verlag für Bauwesen, Berlin (Ost), 388 pp

Ehrenberger R (1931) Direkte Geschiebemessungen an der Donau bei Wien und deren bisherige Ergebnisse. Die Wasserwirtschaft 34:581–589

Ehrenberger R (1942) Geschiebetrieb und Geschiebefracht der Donau in Wien auf Grund direkter Messungen. Wasserkraft Wasserwirtschaft 37(12):265–269

Einstein A (1926) Die Ursache der Mäanderbildung der Flußläufe und des sogenannten Baerschen Gesetzes. Die Naturwissenschaften, Berlin, 14:223–224

Einstein HA (1942) Formulas for the transportation of bedload. Trans ASCE 107:561–577

Einstein HA (1950) The bed-load function for sediment transporation in open channel flows. US Dep Agric, Soil Conserv Serv, Techn Bull 1026:70 pp

Elliot CM (ed) (1984) River meandering. Proc Conf Rivers '83, New Orleans 1983. Am Soc Civ Eng, ASCE-Publ, New York, 1036 pp

Elliot T (1976) The morphology, magnitude and regime of a Carboniferous fluvial-distributary channel. J Sediment Petrol 46:70–76

Embleton C, Thornes J (1979) Process in geomorphology. E Arnold Ltd, Lond, 436 pp

Emmet WW (1980) A field calibration of the sediment-trapping characteristics of the Helley-Smith bedload sampler. US Geol Surv Prof Pap, 1139:44 pp

Engelmann R (1922) Die Entstehung des Egertales. Abh Geogr Ges Wien, Wien, 12:80 pp

Engelsing H (1981) Die Verwendung photoelektrischer Trübungsmesser zur Schwebstoffmessung. Beitr Hydrol, Sonderh 2:193–210

Engelund F (1974) Flow and bed topography in channel bends. Proc Hydrol Div, J Am Soc Civ Eng, 100(11):1631–1648

Engelund F, Hansen E (1967) A monograph on sediment transport in alluvial streams. Kopenhagen, 62 pp

Engelund F, Skovgaard O (1973) On the origin of meandering and braiding in alluvial streams. J Fluid Mech 57(2):289–302

Erdelyszky Z (1974) Messung des Flußsedimenttransportes mit Hilfe radioaktiver Markierung. ÖWW, Wien, 26(3/4):89–92

Ergenzinger P (1982) Über den Einsatz von Magnettracern zur Messung des Grobgeschiebetransportes. Beiträge zur Geologie der Schweiz – Hydrologie. Bd 28 II:483–491

Ergenzinger P, Conrady J (1982) A new tracer technique for measuring bedload in natural channels. Catena 9:77–80

Ergenzinger P, Custer SG (1983) Determination of bedload transport using naturally magnetic tracers: first experiences at Squaw Creek, Gallatin County, Montana. Water Resour Res 19(1):187–193

Erkek C (1967) Beitrag zur Berechnung des Geschiebetriebes in offenen Gerinnen mit beweglicher Sohle unter besonderer Berücksichtigung der Flachlandflüsse. Mitt Leichtweiß-Inst TH Braunschweig 17:127 pp

Ermeling H (1951) Mäander im flachgeneigten Schichtgestein. Ein Beitrag zum Mäanderproblem am Beispiel des Neckars und seiner Nebenflüsse. Diss Univ Mainz, 82 pp

Ertel H (1963) Fluviale Erosion und Akkumulation (Ein Beitrag zur theoretischen Geomorphologie). Monatsber Dtsch Akad Wiss Berl, Berlin (Ost), 5(8/9):515–518

Ertl O (1939) Die Gestaltungsvorgänge am Saalachsee bei Reichenhall und an anderen Stauräumen in alpinen Gewässern. Dtsch Wasserwirtschaft, Stuttgart, 34(7):289–295

Exner FM (1919) Zur Theorie der Flußmäander. SAkW, Math-nat Kl Abt IIa, Wien, 128:1453–1473

Exner FM (1927) Zur Wirkung der Erddrehung auf Flußläufe. Geogr Ann 9:173–180

Fabre LA (1903) La dissymmetrie de vallées et la loi de De Baer, particulièrement en Gascogne. Geographie 8:291–316

Fabricius F, Müller S (1970) A Buret cylinder for grain-size analysis of silt and clay (with ALGOL-program). Sedimentology 14:39–50

Fahnestock RK (1963) Morphology and hydrology of a glacial stream – White River, Mount Rainier, Washington. Geol Surv Prof Pap 422-A:70 pp

Färber K (1987a) Stochastische Modelle zur Simulation des Transportes nicht kohäsiver Sedimente in offenen Gerinnen. Mitt Inst f Wasserwesen Univ d Bundeswehr, München, 18:115 pp

Färber K (1987b) Stochastische Modelle der Bewegung suspendierter Partikel in turbulenter Strömung. Mitt Inst f Wasserwesen Univ d Bundeswehr, München, 21:229 pp

Fargue O (1868) Étude sur la correlation entre la configuration du lit et la profondeur d'eau dans les rivières à fond mobile. Ann Ponts Chaussées, Paris, 4. Ser, Memoires Doc 174:34–92

Fargue O (1882) La Largeur du Lit Moyen de la Garonne. Ann Ponts Chaussées, Paris, Memoires Doc 53:301–328

Faust KM, Denton RA, Plate EJ (1983) Schichtungseffekte in Flußstauhaltungen: Laborversuche. Wasserwirtschaft 73(11):386–392

Feldmann S, Harris SA, Fairbridge RW (1968) Drainage patterns. In: Fairbridge RW (ed) The encyclopedia of geomorphology. Reinhold, NY, pp 284–291

Feldner H (1903) Die Flußdichte und ihre Bedingtheit im Elbsandsteingebirge. Mitt Ver Erdkunde, Jg 1902, Leipzig, pp 1–55

Felkel K (1969) Die Erosion des Oberrheins zwischen Basel und Karlsruhe. Gas- Wasserfach, München, 110(30):801–810

Felkel K (1970) Ideenstudie über die Möglichkeit der Verhütung von Sohlenerosionen durch Geschiebezufuhr aus der Talaue ins Flußbett, dargestellt am Beispiel des Oberrheins. Mitt Bundesanstalt Wasserbau, Karlsruhe, 30:21–29

Felkel K (1980) Die Geschiebezugabe als flußbauliche Lösung. Mitt Bundesanstalt Wasserbau, Karlsruhe, 47:54 pp

Felkel K (1987) Acht Jahre Geschiebezugabe am Oberrhein. Wasserwirtschaft 77. Jg, H 4:181–185

Felkel K, Störmer H-E (1970) Die Geschiebemeßgeräte an der Rheinsohle beim Pegel Maxau und der Einfluß der Schiffahrt auf den Geschiebetrieb. Die Wasserwirtschaft, Stuttgart, 60(11):357–362

Felkel K, Störmer H-E (1980) Akustische Geschiebemessungen im Oberrhein. Wasserwirtschaft, 70(7/8):257–264

Fels E (1948) Junge Flußverlagerungen des Wardar. Naturw Rundsch, Stuttgart, 1(4):174–176

Ferguson RI (1975) Meander irregularity and wavelength estimation. J Hydrol 26(3–4):315–333

Ferguson RI (1977) Meander migration: equilibrium and change. In: Gregory KJ (ed) River channel changes. J Wiley and Sons, NY, pp 236–248

Ferguson RI, Prestegaard KL, Ashworth PJ (1989) Influence of sand on hydraulics and gravel transport in a braided gravel bed river. Water Resour Res 25(4):635–643

Fischer K (1964) Talverschüttungen, Flußverlegungen und Epigenesen im Vinschgau und seinen Nebentälern. Mitt Geogr Ges München 49:181–200

Fisk HN (1952) Mississippi River Valley geology-relation to river regime. Am Soc Civil Eng Trans 117:667–689

Flint JJ (1980) Tributary arrangements in fluvial systems. Am J Sci 280:26–45

Flohn H (1935) Beiträge zur Problematik der Talmäander. Frankf Geogr Hefte 9(1):96 pp

Fluck R (1925) Die Flußdichte im schweizerisch-französischen Jura. Diss, Univ Basel, 102 pp

Folk RL (1965) Petrology of sedimentary rocks. Univ Texas, Austin, 159 pp

Folk RL, Ward WC (1957) Brazos river bar: a study in the significance of grain size parameters. J Sedim Petrol 27:3–26

Förstner U, Müller G, Reineck HE (1968) Sedimente und Sedimentgefüge des Rheindeltas im Bodensee. N Jahrb Miner Abh 109(1/2):33–62

Forchheimer P (1930) Hydraulik. BG Teubner, Leipzig Berlin, 3. Aufl, 596 pp

Fournier F (1969) Transports solides effectuées par les cours d'eau. Bull IASH 14(3):7–49

Franke PG (1974) Hydraulik für Bauingenieure. de Gruyter, Berlin (Sammlung Göschen, 9004), 349 pp

Frécaut R, Pagney P (1978) Climatologie et hydrologie fluviale à la surface de la terre. Livre I. Cent doc univ soc d'édition d'enseignement superieur réunis, Paris, 221 pp

Friedman GM (1961) Distinction between dune, beach, and river sands from their textural characteristics. J Sediment Petrol, Tulsa/Okl, 31:514–529

Füchtbauer H (1959) Zur Nomenklatur der Sedimentgesteine. Erdöl und Kohle 12:605–613

Füchtbauer H, Müller G (1970) Sedimente und Sedimentgesteine. Sedimentpetrologie, Stuttgart, 2: 726 pp

Führböter A (1979) Strombänke (Großriffel) und Dünen als Stabilisierungsformen. Mitt Leichtweiß-Inst Wasserbau, Braunschweig, 67:155–191

Führböter A (1983) Zur Bildung von makroskopischen Ordnungsstrukturen (Strömungsriffel und Dünen) aus sehr kleinen Zufallsstörungen. Mitt Leichtweiß-Inst Wasserbau, 79:1–51

Galay VJ (1983) Causes of river bed degradation. Water Resour Res 19(5):1057–1090

Gallo G, Rotundi L (1954) Trasporto di materiale alluvionale in seno a correnti idriche. Energia elettrica, Burz J (transl) (1965) Deutsche Fassung. Bayer Landesst Gewkde, München 34 pp

Garbrecht G (1961) Abflußberechnungen für Flüsse und Kanäle. Die Wasserwirtschaft 51(2):40–45; (3):72–77

Garbrecht G (1977) Betrachtungen über die Linienführung von Gerinnen. DVWW, Fortbildungslehrg Gewässerausbau, Rotenburg/Fulda, 2:24 pp

Garbrecht G (1981) Gewässerausbau in der Geschichte. Wasser Boden 33(8):372–380

Garbrecht G (1982) Gewässerausbau in der Geschichte. Wasser Boden, 34(1):10–16

Garde RJ, Ranga-Raju KG (1977) Mechanics of sediment-transport and alluvial stream problems. Wiley Eastern Ltd, New Delhi Bangalore Bombay, 483 pp

Gardner TW (1975) The history of part of the Colorado River and its tributaries: an experimental study. Four Corners Geol Soc Guidebook, Field Conf Canyonlands, 9:87–95

Gast T, Bahner H (1976) Eine automatische Apparatur zur Bestimmung der Feststoff-Konzentration in Gewässern und im Abwasser. Chem Ing Tech 48(5):452–454

Gehrig W (1976) Geschiebetrieb. DVWW, Fortbildungslrg Gewässerausbau, Barsinghausen, 1:20 pp

Gehrig W (1981) Die Berechnung des Geschiebetriebanfanges. Mitt BAW 50:21–39

Gerber E (1957) Das Längsprofil der Alpentäler und die Steilenwanderungstheorie. Peterm Geogr Mitt, Gotha, Erg 262 (Geomorphologische Studien, Machatschek-Festschrift):79–90

Gierloff-Emden HG (1953) Flußbettveränderungen in rezenter Zeit. Erdkunde, Bonn, 7(4):298–306

Gilbert GK (1914) The transportation of debris by running water. US Geol Surv Prof Pap 86:263 pp

Gölz B (1985) Wasserwirtschaft in Albanien. Österr Wasserwirtschaft, Wien, 37(3/4):123–133

Gölz E (1987) Zur Sohlenerosion des Niederrheins. Wasserwirtschaft 77, H 7/8:432–436

Gölz E, Tippner M (1985) Korngrößen, Abrieb und Erosion am Oberrhein. DGM 29(4):115–122

Gönnenwein ML (1931) Untersuchungen über die Flußdichte schwäbischer Landschaften. Erdgesch Landesk Abh Schwaben Franken, Öhringen, 13:1–66

Goudie A, Anderson M, Burt T, Lewin J, Richards K, Whalley B, Worsley P (eds) (1981) Geomorphological techniques. Allen and Unwin, Lond

Grabs W (1981) Beitrag zur Beschreibung von Kolmationserscheinungen in einem organisch belasteten Kleingewässer. Beitr Hydrol, Sonderh 2:293–312

Gradmann R (1959) Süddeutschland. Stuttgart 1931, Darmstadt 1959, unv Nachdr 2:553 pp, Nachdruck 1959: Hermann Gentner Verl Bad Homburg, Orig Ausg: Verl J. Engelhorns Nachf, Stuttgart

Graf WH (1971) Hydraulics of sediment transport. McGraw-Hill Book Company, NY, 513 pp

Graf WH (1979) Eine Methode zur Berechnung des Feststoff-Transports. DVWW, Fortbildungslrg Gewässerausbau, Barsinghausen, 1:20 pp

Graf WL (1981) Channel instability in a braided, sand bed river. Water Resour Res 17(4):1087–1094

Gravelius H (1914) Flußkunde (Grundriß der gesamten Gewässerkunde in 4 Bd). Göschen'sche Verlagshandlung, Berlin Leipzig, Bd 1:179 pp

Grebe H (1976) Betrachtungen über die natürliche Grundrißgestalt von Flüssen. Mitt Leichtweißinst Wasserbau TU Braunschweig 50:195–201

Green CP (reviewer), Schumm SA (ed) (1979) Drainage basin morphology (book review). J Hydrol 41(1–2):192–193

Gregory KJ (1966) Dry valleys and the composition of the drainage net. J Hydrol 4:327–340

Gregory KJ (1976) Drainage networks and climate. In: Derbyshire E (ed) Climatic geomorphology. Macmillan Lond, pp 289–315

Gregory KJ (ed) (1977) River channel changes. J Wiley and Sons, NY, 448 pp

Gregory KJ, Gardiner V (1975) Drainage density and climate. Z Geomorp NF 19:287–298

Gregory KJ, Walling DE (1973) Drainage basin form and process. Halsted Press, NY, 456 pp

Gresswell RK, Huxley A (eds) (1965) Standard encyclopedia of the world's rivers and lakes. Weidenfeld & Nicholson (Educational), Lond, 384 pp

Griffiths GA (1983) Stable-channel design in alluvial rivers. J Hydrol 65:259–270

Grimm F (1968) Das Abflußverhalten in Europa – Typen und regionale Gliederung. Wiss Veröff Inst Länderkde, Leipzig, NF 25/26:18–180

Grove AT (1972) The dissolved and solid load carried by some West African rivers: Senegal, Niger, Benue and Shari. J Hydrol 16:277–300

Grube F, Vladi F, Vollmer T (1976) Erdgeschichtliche Entwicklung des unteren Alstertales. Mitt Geol-Paläont Inst Univ Hamburg 46:43–56

Gruber O (1978) Gewässerkunde und Hydrographie im Bundesstrombauamt. ÖWW, Wien, 30(9/10):198−203

Günter A (1971) Die kritische mittlere Sohlenschubspannung bei Geschiebemischungen unter Berücksichtigung der Deckschichtbildung und der turbulenzbedingten Sohlenschubspannungsschwankungen. Mitt Va Wasserbau Hydrol Glaziologie, ETH Zürich, 3:70 pp

Gustafson GC (1973) Quantitative Untersuchung zur Morphologie von Flußbecken unter Verwendung von Orthophotomaterial. Münch Geograph Abh 11:162 pp

Gutersohn H (1932) Relief und Flußdichte. Diss ETH, Zürich, 91 pp

Guy HP, Norman VW (1970) Field methods for measurement of fluvial sediment. Tech Water Resour Invest US Geol Surv, Book 3 (Applications of Hydraulics) Chap C 2, 59 pp

Hack JT (1957) Studies of longitudinal stream profiles in Virginia and Maryland. US Geol Surv Prof Pap 294-B:45−97

Hack JT, Young RS (1959) Intrenched meanders of the North Fork of the Shenandoah River, Virginia. US Geol Surv Prof Pap 354-A:10 pp

Hadley RF, Walling DE (eds) (1984) Erosion and sediment yield: some methods of measurement and modelling. Geo Books, Norwich UK, 218 pp

Hahn HH (1983) Kolloidale Wasserinhaltsstoffe in natürlichen Gewässern. Wasserwirtschaft 73(11):434−441

Hammer T (1972) Stream channel enlargement due to urbanisation. Water Resour Res 8(6):1530−1540

Hampel R (1969) Geschiebewirtschaft in Wildbächen. Die Wasserwirtschaft, Stuttgart, 59(3):64−70

Hampel R (1977) Geschiebewirtschaft in Wildbächen. 1. Teil. Wildbach-Lawinenverbau 41(1):3−34

Hampel R (1978) Geschiebetheorie für die Wildbachverbauung. ÖWW, Wien, 20(11/12):231−238

Hampel R (1980) Die Murenfracht von Katastrophenhochwässern. Wildbach-Lawinenverbau, 44(2):71−102

Hansen E (1967) On the formation of meanders as a stability problem. Coast Engng Lab, Tech Univ Denmark, Basic Res, Progress Report 13:9−13

Hartung F (1959) Ursache und Verhütung der Stauraumverlandung bei Talsperren. Die Wasserwirtschaft, Stuttgart, 49(1):3−13

Hartung F (1973) Stützschwellenkraftwerke. Wasserwirtschaft 63(11/12):349−355

Hartung F (1976) Fluß und Flußbau, ein Rückblick und Ausblick. DVWW, Fortbildungslehrgang Gewässerausbau, Barsinghausen, 1:30 pp

Hartung F (1979) 75 Jahre Nilstau bei Assuan, Entwicklung und Fehlentwicklung. Ber VA Wasserbau TU München, München, 40:37 pp, Bildanhang

Hartung F (1987) Der Assuan-Hochdamm − Fehlplanung oder unvollendet? Wasser und Boden H 9:449−455

Haslam SM (1978) River plants. Cambridge Univ Press, Cambridge, 396 pp

Hayami S (1941) Hydrological studies on the Yangtse River, China. J Shanghai Sci Inst, New Ser, 1(1)

Hayashi T (1970a) Formation of dunes and antidunes in open channels. Proc ASCE, 96(2) Pap 7056:357−366

Hayashi T (1970b) On the cause of the initiations of meandering of rivers. Trans JPN Soc Cic Engrs 2:235−239

Henderson FM (1966) Open channel flow. Macmillan NY, 522 pp

Henkel L (1922) Das Baersche Gesetz dennoch richtig. Peterm Geogr Mitt, Gotha, 68:54−56

Henkel L (1926) Zur Morphologie der Flußläufe. Geol Rdsch 17:1−4

Henkel L (1928) Zum Baerschen Gesetz. Geogr Ann 10:381−382

Hensen W (1939) Der Einfluß der Erdumdrehung auf Tideflüsse in der Natur und im Modell. Die Bautechnik, Berlin, 17(21):285−288

Hensen W (1948) Über den Wert der Kenntnis von der wahren Gestalt einer Flußsohle. Die Wasserwirtschaft, Stuttgart, 39(1):14−17

Herrmann R (1977) Einführung in die Hydrologie. Teubner, Stuttgart, 151 pp

Hey RD (1976) Geometry of river meanders. Nature (Lond) 262:482−484

Hey RD, Thorne CR (1975) Secondary flows in river channels. Area 7:191−195

Hey RD, Bathurst JC, Thorne CR (eds) (1982) Gravel-bed rivers. Wiley Intersci Publ, New York, 875 pp

Hickin EJ (1974) The development of meanders in natural river channels. Am J Sci 274:414−442

Hiersemann D (1973) Die Fluß- und Taldichte im kristallinen Odenwald. Heidelb Geogr Arb 38:125–140

Hinrich H (1965) Beitrag zur Schwebstoffmessung in Wasserläufen mit Beschreibung eines einfachen Filterverfahrens. DGM, Koblenz, 9(3):49–60

Hinrich H (1968) Der Feststofftransport des Rheins, der Weser und der Ems mit Angaben über Meßstellen, Meßgeräte und Auswertung der Meßergebnisse mittels elektronischer Datenverarbeitung. Symp „Aktuelle Fragen der Flußregulierung und der Feststofführung", Budapest, 8.–11.10.68, 7 pp

Hinrich H (1969) Beschreibung eines Querprofilschreibers mit Radiolog und eines Tiefenplanschreibers mit Dopplerlog. DGM, Koblenz, 13(3):73–78

Hinrich H (1971) Schwebstoffgehalt und Schwebstofffracht der Haupt- und einiger Nebenflüsse in der Bundesrepublik Deutschland. Dtsch Gewässerkundl Mitt 15:113–129

Hinrich H (1971) Geschiebemessungen durch Sohlenplatten und Fangbehälter mit eingebautem Hydrophon. Bes Mitt Dtsch Gewk Jahrb, Koblenz, 35:373–378

Hinrich H (1972) Geschiebetrieb und Geschiebefracht des Rheins im Abschnitt Freistett-Worms in den Jahren 1968–1971. DGM, Koblenz, 16(2):29–41

Hinrich H (1973) Der Geschiebetrieb beobachtet mit Unterwasserfernsehkamera und aufgezeichnet durch Unterwasserschallaufnahmegeräte. Wasserwirtschaft, Stuttgart, 63(4):111–114

Hinrich H (1976) Schwebstoffgehalt und Sinkstoffablagerungen in Schiffahrtskanälen. Z Binnenschiffahrt Wasserstraßen 2:43–50

Hinrich H, Schemmer H (1979) Schwebstoffmessungen im Neckar bei Poppenweiler mittels Filtrierautomat und Filterbildgerät. DGM, 23(1):6–9

Hjulström F (1935) Studies of the morphological activity of rivers as illustrated by the river Fyris. Bull Geol Inst Uppsala 25:221–527

Hjulström F (1942) Studien über das Mäanderproblem. Geogr Ann 24:233–269

Hjulström F (1952) The geomorphology of the alluvial outwash plains (Sandurs) of Iceland and the mechanics of braided rivers. Int Congr IGU, 17th Gen Ass, Washington, pp 337–342

Hjulström F, Jonsson J, Sundborg A (1954) The Hoffellssandur. Part I. Geogr Ann, Stockholm, 1–2

Hjulström F, Jonsson J, Sundborg A, Arnborg L (1955) The Hoffellssandur. Part II. Geogr Ann, Stockholm, 3–4

Hoeg S, Voigt G (1967) Über die Haltbarkeit von lumineszierenden Markierungen auf Sandkörnern. Gerlands Beitr Geophys 76:497–512

Hofer B (1987) Der Feststofftransport von Hochgebirgsbächen am Beispiel des Pitzbaches. Österr Wasserwirtschaft. Jg 39, H 1/2:30–38

Hol JBL (1938) Das Problem der Talmäander. Z Geomorph, Berlin, 10:169–195

Holtorff G (1987) Entwicklung natürlicher alluvialer Abflußgerinne. Mitt Franzius-Inst f Wasserbau u Küsteningenieurwesen Univ Hannover, H 65:227–254

Hooke JM (1979) An analysis of the processes of river bank erosion. J Hydrol 42(1–2):39–62

Hormann K (1964) Torrenten in Friaul und die Längsprofilentwicklung auf Schottern. Münchner Geogr Hefte 26:80

Hormann K (1965) Das Längsprofil der Flüsse. Z Geomorph NF 9:437–456

Horton RE (1932) Drainage-basin characteristics. Trans Am Geophys Union 13:350–361

Horton RE (1945) Erosional development of streams and their drainage basins: hydrophysical approach to quantitative morphology. Geol Soc Am Bull 56:275–360

Howard AD (1967) Drainage analysis in geological interpretation. Bull Am Assoc Petrol Geologists 51:2246–2259

Howard AD (1971) Optimal angles of stream junction: geometric stability to capture, and minimum power criteria. Water Resour Res 7(4):863–873

Howard AD, Knutson TR (1984) Sufficient conditions for river meandering: a simulation approach. Water Resour Res 20(11):1659–1667

Hunt CB (1969) Geological history of the Colorado River. US Geol Surv Prof Pap 669:59–130

Hubbel W (1964) Apparatus and techniques for measuring bedload. Geol Surv Water-Supply Pap 1748:74 pp

Hoyt WA, Langbein WB (1955) Floods. Princeton Univ Press, Princeton NJ, 469 pp

Huppmann K (1976) Untersuchung über die Anwendung von Natursteinsicherungen in Verbindung mit Kunststoffiltermatten an der unteren Isar zwischen Gottfrieding und der Mündung in die Donau. Wasserwirtschaftsamt Landshut Bericht, 60 pp (unpublished)

IHP, OHP (1983) Hydrologische Untersuchungsgebiete in der Bundesrepublik Deutschland. Mitt Natl Komm BRD IHP UNESCO OHP WMO, Koblenz, 4:323 pp

Inglis CC (1949) The behaviour and control of rivers and canals. Res Publ Poona, 2 vols, no 13

Jäggi M (1978) Die Sedimenttransportformeln von Meyer-Peter, Einstein und Engelund – Vergleich, Gültigkeitsbereiche, praktische Anwendung. VA Wasserbau, Hydrologie Glaziologie, ETH Zürich, 4:88 pp

Jäggi M (1983) Alternierende Kiesbänke. Mitt VA Wasserbau, Hydrologie Glaziologie, Zürich, 62:286 pp

Jäggi M (1984) Abflußberechnung in kiesführenden Flüssen. Wasserwirtschaft 74(5):263–267

James WR, Krumbein WC (1969) Frequency distributions of stream link lengths. J Geol 77:544–565

Jansen PPH, Van Bendegom L, Van den Berg J, De Vries M, Zanen A (eds) (1979) Principles of river engineering, the non-tidal river. Pitman, Lond San Francisco Melbourne, 509 pp

Jarocki W (1963) A study of sediment (translated from Polish 1957). Nat Sci Found and US Dept of Int, Wash DC

Jarvis RS (1972) New measure of the topologic structure of dendritic drainage networks. Water Resour Res 8:1265–1271

Jarvis RS (1976) Stream orientation structures in drainage networks. J Geol 84:563–582

Jarvis RS, Sham CH (1981) Drainage network structure and the diameter-magnitude relation. Water Resour Res 17(4):1019–1027

Jarvis RS, Werrity A (1975) Some comments on testing random topology stream network models. Water Resour Res 11(2):309–318

Jarvis RS, Woldenberg MJ (eds) (1984) River networks. Benchmark Pap Geol 80:386 pp

Jenks WN (1975) Some considerations of straightness in channel form relating to the estuary of the McLennan River, South Otago (New Zealand). J Hydrol 24(1–2):89–109

Kalweit H (1976) Auswirkungen der Urbanisierung auf die Wasserwirtschaft eines großen Flußgebietes – Modell Rhein. Wasserwirtschaft 66(1/2):14–24

Kalweit H (1976) Flüsse als charakteristische Einzelwesen. DVWW, Fortbildungslrg Gewässerausbau, Barsinghausen, 1:6 pp

Karantounias G (1980) Beitrag zur Hydraulik der Bodenerosion beim Dünnschichtabfluß. Wasserwirtschaft 70(10):347–350

Karl J (1965) Über Auflandungswinkel hinter Geschieberückhaltesperren. Wasser Boden 12:407–409

Karl J (1970) Über die Bedeutung quartärer Sedimente in Wildbachgebieten. Wasser Boden, 9:271–272

Karl J, Höltl W (1974) Analyse alpiner Landschaften in einem homogenen Rasterfeld. Schriftenr Bayer Landesst Gewk, München, 10:32 pp

Karl J, Mangelsdorf J (1971) Typen des fluviatilen Abtrags in den nördlichen Ostalpen. Interpraevent-Tagung 1971, Villach (Austria), Bd 1, pp 23–33

Karl J, Mangelsdorf J (1975) Die Wildbäche der Ostalpen. Interpraevent Innsbruck, Tagungspubl 1:397–406

Karl J, Mangelsdorf J, Scheurmann K (1975) Der Geschiebehaushalt eines Wildbachsystems, dargestellt am Beispiel der oberen Ammer. DGM, Koblenz, 19(5):121–132

Karl J, Mangelsdorf J, Scheurmann K (1977) Die Isar – ein Gebirgsfluß im Spannungsfeld zwischen Natur und Zivilisation. Jb Ver Schutz Bergwelt, München, 42:175–224

Kashef A (1981) The Nile – one river and nine countries. J Hydrol 53(1–2):53–71

Kaufmann H (1929) Rhythmische Phänomene der Erdoberfläche. Braunschweig

Keller EA (1972) Development of alluvial stream channels: a five stage model. Geol Soc Am Bull 83:1531–1536

Keller R (1962) Gewässer und Wasserhaushalt des Festlandes. Teubner, Leipzig, 520 pp

Keller R, Skirke S, Seifried A (1972) Methoden zur Klassifikation von Abflußregimen, 2. Ber d IGU-Kommission in der IHD, engl u deutsch. Freiburger Geogr Hefte 12:179 pp

Kelsey HM, Lamberson R, Madej MA (1987) Stochastic model for the long-term transport of stored sediment in a river channel. Water Resour Res 23(9):1738–1750

Kotoulas K (1986) Natürliche Entwicklung der Längen- und Querprofilform der Flüsse – ein Beitrag zum naturnahen Flußbau. Veröff Inst f Siedlungs- und Industriewasserbau TU Graz, H 12:259 pp

Kempe S, Mycke B, Seeger M (1981) Flußfrachten und Erosionsraten in Mitteleuropa, 1966–1973. Wasser Boden 33(3):126–131

Kennedy JF (1963) The mechanics of dunes and antidunes in erodible-bed channels. J Fluid Mech 16(4):521–544

Kirkby MJ, Chorley RJ (1967) Throughflow, overland flow and erosion. IASH Bull 10:74–85

Klapper H (Hrsg) (1980) Flüsse und Seen der Erde. H Klapper, Leipzig Jena Berlin, 240 pp

Kleinschroth A, Abelein R (1985) Modelluntersuchungen über die Wasserspiegellage in der Krümmung einer Schußrinne. Hydraulik Gewässerkde TU München, München, 44:199–224

Klostermann H (1970a) Die genetische Einteilung der Regimefaktoren. WWT 20(9):297–299

Klostermann H (1970b) Zur geomorphometrischen Kennzeichnung kleiner Einzugsgebiete. Peterm Mitt, Gotha, 104(4):241–260

Knauss J (1977) Flachgeneigte Abstürze, glatte und rauhe Sohlrampen. DVWW-Fortbildungslrg für Gewässerausbau, Rotenburg/Fulda, 2:44 pp

Knighton AD (1984) Indices of flow asymmetry in natural streams: definition and performance. J Hydrol 73:1–19

Köhn M (1929) Korngrößenbestimmung vermittels Pipettanalyse. Tonind Z Keram Rundsch 53:729–731

König G (1971) Flußordnungsanalyse im Gebiet zwischen Isar und Inn. Unveröff Zulassungsarbeit, Geogr Inst Univ München

Köster E (1964) Granulometrische und morphometrische Meßmethoden. F Enke, Stuttgart, 336 pp

Komura S, Simons DB (1967) River-bed degradations below dams. J Hydraulics Div, Proc ASCE, Hy 4

Kothé P (1967) Die Biologie als Hilfsmittel bei der Erforschung morphologisch-quantitativer Vorgänge in den Gewässern. DGM, Koblenz, Sonderh:220–228

Kremer E (1954) Die Terrassenlandschaft der mittleren Mosel. Ein Beitrag zur Quartärgeschichte. Arbeit Rhein Landesk, Bonn, 6:100 pp

Kreps H (1961) Das Problem der Ermittlung einer Jahresgeschiebefracht aus einer beschränkten Anzahl von Messungen. ÖWW, Wien, 13(5/6):96–99

Kresser W (1964) Gedanken zur Geschiebe- und Schwebstofführung der Gewässer. ÖWW, Wien, 16(1/2):6–11

Kresser W (1967) Die historische Entwicklung der Gewässerkunde und ihre Bedeutung in unserer Zeit. DGM Sonderh, pp 5–11, 231 pp

Kresser W (1973) Die Donau und ihre Hydrologie. Wasser-Energiewirtschaft Sonderh März/April: 83–99

Krier H (1983) Erosionsbeginn bei kohäsiver Wasserlaufsohle. Wasserbau-Mitt, TH Darmstadt, 22:75–107

Krigström A (1962) Geomorphological studies of sandur plains and their braided rivers in Iceland. Geograf Ann 34:328–346

Kronberg P (1984) Photogeologie. F Enke, Stuttgart, 268 pp

Kuenen P (1953) Origin and classification of submarine canyons. Bull Geol Soc Am 64:1295–1314

Kuhnle RA, Southard JB (1988) Bed load transport fluctuations in a gravel bed laboratory channel. Water Resour Res 24(2):247–260

Kuhr HH (1972) Die Länge des Rheins und seine Vermessung. Beitr Rheinkde 24(2):3–15

Kurzmann E (1953) Über die strömungstechnischen Grundlagen der Schwebstoffprobleme. ÖWW, Wien, 5(11):221–227

Kurzmann E (1964) Die bettformenden Kräfte an Stauraumufern. ÖWW, Wien, 16(1/2):12–15

Lammel K (1986) Berechnung der Vertikalstruktur dichteabhängiger Strömungsvorgänge im Mündungsbereich eines Seezuflusses. Hydraulik u Gewässerkde, TU München, Mitt Nr 46: 105 pp

Lane EW (1955) The importance of fluvial morphology in hydraulic engineering. Proc ASCE, 81(745):17 pp

Lane EW, Kalinske HA (1939) The relation of suspended to bed load materials in rivers. Transactions Am Geophys Union 4:637

Langbein W, Leopold LB (1966) River meanders – theory of minimum variance. US Geol Surv Prof Pap 422-H, pp 1–15

Langbein WB, Schumm SA (1958) Yield of sediment in relation to mean annual precipitation. Trans Am Geophys Union 39:1076–1084

Lanser O (1953) Die bisherige Entwicklung der Geschiebetheorien und Geschiebebeobachtungen. Blätter Technikgeschichte, Wien, H 15 (Gedenkschrift Sechzig Jahre Hydrographischer Dienst in Ö):58–78

Lauterborn R (1916–1918) Die geographische und biologische Gliederung des Rheinstroms. Sitz Ber Heidelberger Akad Wiss, Math-nat Kl Abt B, I: 61 pp, 1916; II: 70 pp, 1917; III: 87 pp, 1918

Leder A (1970) Statistische Flußgebietsuntersuchungen nach geomorphologischen Gesichtspunkten für hydrologische Zwecke im Gebiet der DDR. Diss Humboldt Univ Berlin, Berlin, 106 S, 33 Anl

Lee LJ, Henson BL (1977) The interrelationships of the longitudinal profiles and channel pattern for the Red River. J Hydrol 35:191–201

Leeder MR (1982) Sedimentology. Allen and Unwin, Lond, 344 pp

Leliavsky S (1966) An introduction to fluvial hydraulics. Dover Publ, NY

Lemcke K, Von Engelhardt W, Füchtbauer H (1953) Geologische und sedimentpetrographische Untersuchungen im Westteil der ungefalteten Molasse des Süddeutschen Alpenvorlandes. Beih Geol Jb, Hannover, 11:182 pp

Leopold LB (1962) Rivers. Am Sci 50(4):511–537

Leopold LB (1970) An improved method for size distribution of streambed gravel. Water Resour Res 6(5):1357–1366

Leopold LB (1973) River channel change with time – an example. Bull Geol Soc Am 84:1845–1860

Leopold LB (1976) Reversal of erosion cycle and climatic change. Quat Res 6:557–562

Leopold LB, Langbein WB (1966) River meanders. Sci Am 6(June):60–77

Leopold LB, Miller JP (1956) Ephemeral streams: hydraulic factors and their relation to the drainage net. US Geol Surv Prof Pap 282-A:36 pp

Leopold LB, Wolman MG (1957) River channel patterns; braided, meandering and straight. US Geol Surv Prof Pap 282:39–85

Leopold LB, Wolman MG (1960) River meanders. Bull Geol Soc Am 71:769–794

Leopold LB, Wolman MG, Miller JP (1964) Fluvial processes in geomorphology. WH Freeman & Co, San Francisco Lond, 522 pp

Lewin J (1976) Initiation of bed forms and meanders in coarse grained sediment. Geol Soc Am Bull 87:281–285

Lichtenhahn C (1977 a) Flußbau. Erweiterte Vorlesung ETH Zürich, 1:244 pp, 2:274 pp

Lichtenhahn C (1977 b) Bestimmung und Sicherung des Regelungslängsschnittes. Vortr Fortbildungslehrg „Gewässerausbau" DVWW, Rotenburg/Fulda, 2:45 pp

Limerinos JT (1970) Determination of the Manning coefficient from measured bed roughness in natural channels. US Geol Surv Water Supply Pap 1898-B:47 pp

List FK (1969) Quantitative Erfassung von Kluftnetz und Entwässerungsnetz aus dem Luftbild. Bildmess Luftbildwesen, Karlsruhe, 37:134–140

List FK, Helmcke D (1970) Photogeologische Untersuchungen über lithologische und tektonische Kontrolle von Entwässerungssystemen im Tibesti-Gebirge, Zentral-Sahara, Tschad. Bildmess Luftbildwesen, Karlsruhe, 38(5):273–278

List FK, Stock P (1969) Photogeologische Untersuchungen über Bruchtektonik und Entwässerungsnetz im Präkambrium des nördlichen Tibesti-Gebirges, Zentral-Sahara, Tschad. Geol Rdsch, Stuttgart, 59(1):228–256

Lopatin GW (1962) Intensität der Wassererosion auf dem Gebiet der UdSSR. Symp IAHS Comm Land Erosion, Bari, IAHS 59:43–51

Louis H (1961) Allgemeine Geomorphologie. Lehrb Allg Geogr, Berlin, 1:355 pp

Louis H, Fischer K (1979) Allgemeine Geomorphologie. de Gruyter, Berlin, 4. Aufl, 814 pp

Louis H (Hrsg) (1971) Landformung durch Flüsse. Z Geomorphologie (Suppl) 12:237 pp

Louis H (1975) Abtragungshohlformen mit konvergierend-linearem Abflußsystem. Zur Theorie des fluvialen Abtragungsreliefs. Münchner Geogr Abh, München, 17:45 pp

Lubowe JK (1964) Stream junction angles in the dendritic drainage pattern. Am J Sci 262:325–339

Machatschek F (1943) PS Jovanovics Untersuchungen über das Längsprofil von Flüssen. Peterm Geogr Mitt, Gotha, 89(3/4):102–104

Machatschek F (1953) Über seitliche Erosion. Peterm Geogr Mitt, Gotha, 97:24–26

Machatschek F (1964) Geomorphologie. Teubner, Stuttgart, 8. Aufl (bearb Graul, Rathjens), 209 S

Mackin JH (1948) Concept of the graded river. Geol Soc Am Bull 59:463–512

Madduma Bandara CM (1974) Drainage density and effective precipitation. J Hydrol 21(2):187–190

Mahard RH (1942) The origin and significance of intrenched meanders. J Geomorph 5:32–44

Maniak U (1967) Geschiebe- und Schwebstofführung der Oker. Mitt Leichtweiß-Inst TH Braunschweig, 20 (Zimmermann-Festschrift):145−167

Marcinek J (1975) Das Wasser des Festlandes. H Haack, Leipzig 224 pp

Mark DM (1974) Line intersection method for estimating drainage density. Geology, 2:235−236

Marsal D (1979) Statistische Methoden für Erdwissenschaftler. Schweizerbart, 2. Aufl, 192 pp

Martinec J (1967) Berechnung der Geschwindigkeit unter Berücksichtigung der meßbaren Flußgerinneparameter. Die Wasserwirtschaft 57(11):395−398

Martinec J (1972) Comment on the paper "On river meanders" by Ch T Yang. J Hydrol, 15(3): 249−251

Masuch K (1935) Zur Frage der Talmäander. Berliner Geogr Arb 9:46 pp

Maull O (1958) Handbuch der Geomorphologie. F Deuticke, Wien, 2. Aufl, 600 pp

Maxwell JC (1955) The bifurcation ratio in Horton's law of stream numbers. Am Geophys Union Trans, Washington, Abstr, 36(3):520 ff

Mayrhofer A (1963) Normalprofil und Ausgleichsgefälle. Diss TH Wien, Wien, 74 pp

Mayrhofer A (1964) Normalprofil und Ausgleichsgefälle geschiebeführender Flüsse. ÖWW, Wien, 16(7/8):149−156

Mayrhofer A (1970a) Über die Gesetzmäßigkeiten des Geschiebetriebes. ÖWW, Wien, 22(5/6): 170−184

Mayrhofer A (1970b) Wirkung von Sohlgurten und Sohlschwellen bei Flußregulierungen. Ber Flußbautagung Bregenz, 9: pp 112−137, 186 pp

McArthur DS, Ehrlich R (1977) An efficiency evaluation of four drainage basin shape parameters. Prof Geogr, 29(3):290−295

Melton MA (1958) Geometric properties of mature drainage systems and their representation in an E_4 phase space. J Geol, Chicago, 66:35−56

Melzer KJ (1976) Das Hochwasserschutzsystem am unteren Mississippi und die Überschwemmung des Jahres 1973. Die Bautechnik 53(5):168−173

Mensching H (1957) Bodenerosion und Auelehmbildung in Deutschland. DGM, Koblenz, 1(6):110−114

Mertens W (1977) Die natürliche Grundrissgestalt des ungeregelten Flusses. Vortr Fortbildungslrg „Gewässerausbau" DVWW Rotenburg/Fulda, 2:20 pp

Mertens W (1987) Über die Deltabildung in Stauräumen. Mitt Leichtweiß-Inst H 91:1−149

Mesa OJ, Gupta VK (1988) On the main channel length-area relationship for channel networks. Water Resour Res 23(11):2119−2122

Metz B (1981) Ursachen und zeitliche Einordnung der Souris River Umlenkung in Südmanitoba (Kanada). Beitr Hydrol Sonderh 2:147−160

Meyer-Peter E, Favre H, Einstein A (1934) Neuere Versuchsresultate über den Geschiebetrieb. Schweizer Bauz 13, Bd 103:147−150

Meyer-Peter E, Favre H, Müller R (1935) Beitrag zur Berechnung der Geschiebeführung und der Normalprofilbreite von Gebirgsflüssen. Schweizer Bauz 9/10 Bd 105:95−99, 109−113

Meyer-Peter E, Müller R (1949) Eine Formel zur Berechnung des Geschiebetriebs. Schweiz Bauz, Zürich, 67(3):4 pp

Meyer-Peter E, Lichtenhahn C (1963) Altes und Neueres über den Flußbau. Veröff Eidgen Amtes Straßen- Flußbau, Bern, 55 pp

Miall AD (1977) Fluvial sedimentology: Fluvial lecture series notes. Can Soc Petrol Geol, 111 pp

Miall AD (ed) (1978) Fluvial sedimentology. Can Soc Petrol Geol Memoir, Calgary, 5:859 pp

Milne JA (1982a) Bend forms and bend-arc spacings of some course-bedload channels in upland Britain. Earth Surf Proc Landforms 7:227−240

Milne JA (1982b) Bed-material size and the riffle-pool sequence. Sedimentology 29:267−278

Milton LE (1967) An analysis of the laws of drainage net composition. Bull Int Assoc Sci Hydrol 12(4):51−56

Mock SJ (1971) A classification of channel links in stream networks. Water Resour Res 7:1558−1566

Moore RC (1926) Origin of inclosed meanders on streams of the Colorado plateau. J Geol 34:29−57

Morawetz S (1941) Zur Mäanderfrage. Peterm Geogr Mitt, Gotha, 87(7/8):263−267

Morgan RPC (1973) The influence of scale in climatic geomorphology: a case study of drainage density in West Malaysia. Geogr Ann 55A:107−115

Morisawa ME (1957) Accuracy of determination of stream lengths from topographic maps. Trans Am Geophys Union 38:86−88

Morisawa ME (1958) Measurement of drainage basin outline form. J Geol 66(5):587–591

Morisawa ME (1962) Quantitative geomorphology of some watersheds in the Appalachian Plateau. Geol Soc Am Bull 73:1025–1046

Morisawa ME (1963) Distribution of stream-flow direction in drainage patterns. J Geol 71(4):528–529

Morisawa M (1968) Streams, their dynamics and morphology. McGraw-Hill (Earth and Planetary Science Ser) New York, 175 pp

Morisawa M (ed) (1973) Fluvial geomorphology. Publ Geomorph, State Univ NY, Binghampton NY, 314 pp

Morisawa M (1985) Rivers. In: Clayton KM (ed) Geomorphology texts 7. Longman, Lond, 222 pp

Mortensen H (1942) Zur Theorie der Flußerosion. Nachr Akad Wiss Göttingen, Math-phys Kl: 35–56 pp

Mortensen H, Hövermann J (1957) Filmaufnahmen der Schotterbewegungen im Wildbach. Peterm Geogr Mitt, Gotha, Ergh 262 (Machatschek-Festschrift):43–52

Moser M (1971) Zahl, Form, Vorgang und Ursache der Anbruchsbildung und ihre Beziehungen zum geologischen Untergrund im Bereich des mittleren Lesachtales (Kärnten). Tagungspubl Interpraevent 1:35–48

Moser M (1973) Analyse der Anbruchsbildung bei den Hochwasserkatastrophen der Jahre 1965 und 1966 im mittleren Lesachtal (Kärnten). Carinthia II, Klagenfurt, 83:179–234

Mosley MP (1976) An experimental study of channel confluences. J Geol 84:535–562

Moss AJ (1962/63) The physical nature of common sandy and pebbly deposits. Parts I and II. Am J Sci 260(5):337–373; 261(4):297–343

Mostafa MG (1955) River-bed degradation below large capacity reservoirs. ASCE Transactions Pap 2879:688–704

Mühlhofer L (1933) Schwebstoff- und Geschiebemessungen am Inn bei Kirchbichl (Tirol). Wasserkraft Wasserwirtschaft 28(4):37–41

Mühlhofer L (1955) Eine Geschiebetriebformel auf theoretischer Grundlage. ÖWW, Wien, 7(5/6):114–119

Müller G (1964) Methoden der Sedimentuntersuchung. Schweizerbart, Stuttgart, 303 pp

Müller G (1966) The New Rhine Delta in Lake Constance. In: Shirley ML, Ragsdale (eds) Deltas in their geologic framework. Geo Soc Houston, pp 107–124

Müller J (1976) Zur Methodik von Schwebstoffuntersuchungen an Flußwässern. gwf-Wasser/Abwasser 117(5):220–222

Müller R (1943) Theoretische Grundlagen der Fluß- und Wildbachverbauungen. Diss, ETH Zürich, 193 pp

Müller R (1960) Die Entwicklung der flußbaulichen Hydraulik. Wasser-Energiewirtschaft, 52(8/9/10): 292–300

Müller S (1973) Hydrogeologische und hydrologische Untersuchungen in der Pupplinger Au im Isartal südlich von München. Diss TU München

Mundschenk H (1979) Zur Quantifizierung von Sedimentbewegungen im Bereich alternierender Tideströmungen. Dtsch Gewässerkundl Mitt 23(5):122–137

Mundt G (1959) Die Untersuchung des Einflusses der Geschiebe- und Schwebstofführung auf die Sohlen- und Wasserspiegellage in Stauseen, abgeleitet aus Messungen und Beobachtungen an der Innstufe Ering. Inst Wasserbau TU Berlin, Berlin, 51:77 S

Murawski H (1964) Kluftnetz und Gewässernetz. N Jb Geol Paläont, Stuttgart, Mh:537–561

Murawski H (1983) Geologisches Wörterbuch. F Enke, Stuttgart, 7. Aufl, 280 pp, 8. Aufl, 281 pp

Muth W (1977) Ermittlung des zulässigen Gefälles zwischen Sohlenbauwerken. Wasser Boden 29(8):225–228

Müürsepp P (1987) Warum die Flüsse das rechte Ufer unterspülen. Das Baer-Babinetsche Gesetz. Geowiss i u Zeit, 5. Jg, 3:102–106

Nace RL (1974) History of hydrology – a brief summary. Nature Resources 10(3):2–9

Nag SK (1960) Berechnung der Bewegung von Schwebstoffen mittels der Potentialtheorie. Diss TH Karlsruhe, 45 pp

Nakel E (1961) Wo stehen wir heute in der Geschiebeforschung? WWT 11(12):598–602

Nasner H (1974) Über das Verhalten von Transportkörpern im Tidegebiet. Mitt Franzius-Inst Wasserbau, TU Hannover, 40:1–149

Nasner H (1984) Wellenerzeugter Sedimenttransport. Mitt Franzius-Inst Wasserbau, TU Hannover, 59:327–339

Neill CR (1968) Bed forms in the lower Red Deer River, Alberta. J Hydrol 7(1):58–85

Neumann L (1900) Die Dichte des Flußnetzes im Schwarzwalde. Gerlands Beitr Geophys, Leipzig, 4:219–240

Neumann R (1963) Die Auswertung von Korngrößenverteilungen durch Häufigkeitsanalyse. N Jb Geol Pal, Mh 9:492–501

Nippes K-R (1982) Erfassung des Schwebstofftransportes in Mittelgebirgsflüssen. 14. DVWK-Fortbildungslehrgang Hydrologie, Hydrometrie, Okt 1982 Andernach, 23 pp

Nordin CF (1971) Statistical properties of dune profiles; sediment transport in all channels. US Geol Surv Prof Pap, 562-F:41 pp

Nordin CF, Rathbun RE (1970) Field studies of sediment movement using fluorescent tracers. Int Symp Hydrometry, Koblenz, Bes Mitt DGM 35:1–12

Nordin CR, Richardson EV (1968) Statistical descriptions of sand waves from streambed profiles. Int Assoc Sci Hydrol Bull 13(3):25–32

Nordseth K (1972) Fluvial processes and adjustments on a braided river. The islands of Koppangsöyene on the River Glomma. Norsk Geogr Tidsskr 27:77–108

Odgaard AJ (1987) Streambank erosion along two rivers in Iowa. Water Resour Res 23(7):1225–1236

Odgaard AJ, Bergs MA (1988) Flow processes in a curved alluvial channel. Water Resour Res 24(1):45–56

Oexle L (1936) Die Schwebstoff- oder Schlammführung der geschiebeführenden Flüsse in Bayern. Wasserkraft Wasserwirtschaft, München, 31(11):20 pp

O'Neill MP, Abrahams D (1984) Objective identification of pools and riffles. Water Resour Res 20(7):921–926

O'Neill MP, Abrahams AD (1986) Objective identification of meanders and bends. J Hydrol 83:337–353

Onesti LJ, Miller TK (1974) Patterns of variation in a fluvial system. Water Resour Res 10(6): 1178–1186

Onesti LJ, Miller TK (1978) Topological classifications of drainage networks: an evaluation. Water Resour Res 14(1):144–148

Ongley ED (1968) An analysis of the meandering tendency of Serpentine Cave, NSW. J Hydrol 6(1):15–31

Ongley ED (1970) Drainage basin axial and shape parameters from moment measures. Can Geogr 14(1):38–44

Otremba E (1950) Der Unterlauf der Flüsse im Luftbild. Peterm Geogr Mitt, Gotha, 94:177–178

Paluska A, Degens ET (1976) Landschaftliche Veränderungen des unteren Alstertales – ein Abbild der Klimaentwicklung seit der letzten Eiszeit. Mitt Geol-Paläont Inst Univ Hamburg 46:5–14

Pardé M (1947) Les méandres des rivières. Rev Geogr Pyrénées Sud-Ouest, Toulouse, 16/17:67–88

Pardé M (1964) Fleuves et Rivières. Paris, 4. edn, Collection A Colin, 223 pp

Park CC (1977) World-wide variations in hydraulic geometry exponents of stream channels: an analysis and some observations. J Hydrol 33:133–146

Parker G, Andrews ED (1985) Sorting of bed load sediment by flow in meander bends. Water Resour Res 21(9):1361–1373

Parvis M (1950) Drainage pattern significance in airphoto identification of soils and bed rocks. Highway Res Board Bull 28:387–400

Paschinger V (1957) Die Flußdichte der Schobergruppe in regionaler Betrachtung. Mitt Geogr Ges Wien, Wien, 99(213):187–193

Paul W (1951) Die Mechanik der Flußablenkungen im Grundgebirge und im Deckgebirge des Südschwarzwaldes. Mitt Bad Geol LA, Freiburg, 1950, pp 115–120

Penck A (1894) Morphologie der Erdoberfläche. Engelhorn, Stuttgart, 1:471 pp, 2:696 pp

Penck A, Brückner E (1909) Die Alpen im Eiszeitalter. Chr Herm Tauchnitz, Leipzig, 3 Bde

Pernecker L, Vollmers H (1965) Neue Betrachtungsmöglichkeiten des Feststofftransportes in offenen Gerinnen. Die Wasserwirtschaft, Stuttgart, 55(12):386–391

Pettijohn FJ (1975) Sedimentary rocks. Harper and Row, NY, 2nd edn 718 pp, 3rd end 628 pp

Pettijohn FJ, Potter PE, Siever R (1972) Sand and sandstone. Springer, Berlin Heidelberg New York, 618 pp

Philippson A (1924) Grundzüge der Allgemeinen Geographie. Akad Verlagsges mbH, Leipzig, Bd 2, p 97 f, 2. Aufl 1930/31

Philippson A (1947) Zur Theorie der Flußerosion. Erdkunde, Bonn, 1(4/6):212−213

Philips PJ, Harlin JM (1984) Spatial dependency of hydraulic geometry exponents in a subalpine stream. J Hydrol 71:277−283

Picard MD, High LR (1973) Sedimentary structures of ephemeral streams. Dev Sedimentology, Elsevier Amst, 17:215 pp

Pichl K (1960) Die Erosion der Rheinsohle zwischen Oppenheim und Koblenz. DGM, Koblenz, 4(5):97−105

Pickup G (1976) Adjustment of stream-channel shape to hydrologic regime. J Hydrol 30(4):365−373

Pickup G, Higgins RJ (1979) Estimating sediment transport in a braided gravel channel; the Kawerong River, Bougainville, Papua New Guinea. J Hydrol 40(3−4):283−297

Pickup G, Higgins RJ, Grant I (1983) Modelling sediment transport as a moving wave − the transfer and deposition of mining waste. J Hydrol 60:281−301

Pitlick J (1988) Variability of bed load measurement. Water Resour Res 24(1):173−177

Pizzuto JE (1984) An evaluation of methods for calculating the concentration of suspended bed material in rivers. Water Resour Res 20(10):1381−1389

Plate E (1974) Hydraulik zweidimensionaler Dichteströmungen. Mitt Inst Wasserbau, Univ Karlsruhe, 3(3):162 pp

Poser H, Müller T (1951) Studien an den asymmetrischen Tälern des Niederbayerischen Hügellandes. Nachr Akad Wiss Göttingen, Math-phys Kl IIb, Jahrg 1951:1−32

Potter PE, Pettijohn FJ (1963) Paleocurrents and basin analysis. Springer, Berlin Göttingen Heidelberg, 296 pp

Press H, Jambor F, Leopold E (1953) Verteilung der Geschiebeführung eines Wasserlaufes, der sich in mehrere natürliche oder künstliche Arme teilt. Dtsch Ber XVIII Intern Schiffahrtskongr Rom, pp 78−97

Press H, Schröder R (1966) Hydromechanik im Wasserbau. W Ernst u Sohn, Berlin München, 548 pp

Pritchard DW (1967) What is an estuary? Physical viewpoint. In: Lauff GH (ed) Estuaries. Am Acad Adv Sci 83:3−5

Prowald HG (1973) Die Geschiebebewegung im kanalisierten Neckar. Heidelb Geogr Arb 38:66−86

Puls E (1910) Vergleichende Untersuchungen über Flußdichte. Diss Univ Hamburg, Hamburg, 40 pp

Puls W (1983) Mathematisches Modell für den Feststofftransport über Sohlformen. Wasserwirtschaft 73(5):143−148

Putzinger J (1919) Das Ausgleichsgefälle geschiebeführender Wasserläufe und Flüsse. Z Österr Ing Arch Ver 13:119−123

Putzinger J (1927) Das Ausgleichs- oder Kompensationsprofil. Z Geschiebef, Bd III, 1/2:82−96

Putzinger J (1954/55) Über die Ursachen des Schlingerns der Flüsse. Geol Bauwesen, Wien, 21(3):97−109

Quick MC (1974) A mechanism of meandering. Proc Hydrol Div, J Am Soc Civ Eng 100(6):741−753

Quitzow HW (1976) Die erdgeschichtliche Entwicklung des Rheintales. Natur Museum 106(11):339−342

Ramesh R, Subramanian V (1988) Temporal, spatial and size variation in the sediment transport in the Krishna River Basin, India. J Hydrol 98:53−65

Ramsay AC (1874) On the physical history of the Rhine. Proc R Inst Gt Britain 7:279−288

Ranalli G, Scheidegger AE (1968) A test of the topological structure of river nets. Bull Int Assoc Sci Hydrol 13(2):142−153

Range W (1961) Morphometrische Untersuchungen in den Einzugsgebieten der bayerischen Alpenflüsse. Veröff Arbeitsber Bayer Landesst Gewkde, München, 51 pp

Rasehorn F (1913) Die Flußdichte im Harz und seinem nördlichen Vorlande. Z Gewk, Dresden, 11:1−56

Rassl G (1967) Untersuchung des Geschiebetransports im Rhein mit Hilfe von radioaktiv markiertem Sand. DGM Sonderh, Koblenz, pp 142−150, 231 pp

Raudkivi AJ (1963) Study of sediment ripple formation. Proc ASCE, 89(6) pap 3692:15−33

Raudkivi AJ (1976) Loose boundary hydraulics. Pergamon Press, Oxford, 2nd edn

Rehbock T (1929) Bettbildung, Abfluß und Geschiebebewegung bei Wasserläufen. Z Dtsch Geol Ges 81:497−534

Reindl C (1939) Leonardo da Vinci, der Hydrauliker. Wasserkraft Wasserwirtschaft 34(23/24): 267–269

Reineck HE (1967) Vergleich von Schlämmanalysen mit der Pipette nach Köhn und dem Areometer nach Casagrande. Senckenb Lethaea 48:351–356

Reineck HE, Singh IB (1980) Depositional sedimentary environments. Springer, Berlin Heidelberg New York, 2nd edn, 549 pp

Reinemann L, Schemmer H, Tippner M (1982) Trübungsmessungen zur Bestimmung des Schwebstoffgehalts. Deutsche Gewässerk Mitt, 26. Jg, H 6:167–174

Remy-Berzenkovich E (1959a) Eine neue Methode zur Ermittlung des Feststofftriebes in Flußläufen. ÖWW, Wien, 11(3):59–66

Remy-Berzenkovich E (1959b) Analyse des Feststofftriebes in Flußläufen. Die Wasserwirtschaft 49(10):1–6

Remy-Berzenkovich E (1960) Analyse des Feststofftriebes fließender Gewässer. Schriftenr Österr WWV, Wien, 41:56 pp

Rhein-Museum eV Koblenz (1962) Die Entstehung des Rheintals vom Austritt des Flusses aus dem Bodensee bis zur Mündung. Beitr Rheinkde, Koblenz, 14:10–47; Wittmann O: Hochrhein und Oberrhein bis Karlsruhe; Wagner W: Der Rhein im Rheintalgraben und im Mainzer Becken; Quitzow HW: Mittelrhein und Niederrhein

Rich JL (1914) Certain types of stream valleys and their meaning. J Geol 22:469–497

Richards KS (1976) Channel width and the riffle-pool sequence. Geol Soc Am Bull 87:883–890

Richards KS (1978) Simulation of flow geometry in a riffle-pool stream. Earth Surf Proc, 3:345–354

Richards KS (1982) Rivers. Univ Paperbacks 771, Lond NY, 358 pp

Rohde H, Meyn G (1972) Untersuchungen über das hydrodynamische Verhalten oberflächenmarkierten Sandes und über die Einbringmethode bei Leitstoffuntersuchungen. Mitt Bundesanst Wasserbau, Karlsruhe, 14 pp

Rohdenburg H (1971) Einführung in die klimagenetische Geomorphologie. Lenz-Verlag, Gießen, 2. Aufl, 350 pp

Rössert R (1976) Grundlagen der Wasserwirtschaft und Gewässerkunde. Oldenbourg München Wien, 2. Aufl, 302 pp

Rössert R (1978) Hydraulik im Wasserbau. Oldenbourg München Wien, 4. Aufl, 180 pp

Rouse HM (1936) Discussion on modern conceptions of the mechanics of fluid turbulence. Trans Am Soc Civ Eng 102:523–543

Rouvé G (ed) (1987) Hydraulische Probleme beim naturnahen Gewässerausbau (Bd 2 d Schwerpunktprogramms: Anthropogene Einflüsse auf hydrologische Prozesse). DFG-Forschungsbericht, 267 pp

Ruck KW (1967) Erfahrungen mit Sandwanderungsuntersuchungen mittels Luminophoren. Die Wasserwirtschaft 10:363–367

Ruck KW (1977) Erfahrungen beim Präparieren von Sand für Leitstoffuntersuchungen. Mitt Bundesanst Wasserbau, Karlsruhe, 33:17–33

Ruck KW (1979) Ingenieurgeologische Naturversuche mit markiertem Geschiebe. Ber Nat Tagung Ingenieurgeol, Fellbach, 2:283–296

Rücklin H (1952) Sohlengeschwindigkeit und Geröllbewegung. Geologie Bauwesen, Wien, 19(3):145–164

Rüdiger A, Leder A, Balek J, Schmidt H (1966) Beiträge zum Abflußvorgang und zur Bettgestaltung in der Elbe. Bes Mitt Gewkd Jahrb DDR, Berlin (Ost), 6:137 pp

Rust BR (1972) Structure and process in a braided river. Sedimentology 18:221–245

Rutte E (1971) Pliopleistozäne Daten zur Änderung der Hnuptabdachung im Main-Gebiet, Süddeutschland. Z Geomorph NF (Suppl) 12:51–72

Rzoska J (1978) On the nature of rivers with case stories of Nile, Zaire and Amazon. Dr W Junk, Den Haag, 67 pp

Samojlov IV (1956) Die Flußmündungen. Haack, Gotha, 1. Aufl, 647 pp

Sandra BC, Machado MB, De Meis MRM (1975) Drainage basin morphometry on deeply weathered bedrocks. Z Geomorph, NF, Berlin, 19:125–139

Sauerbier H (1918) Die Flußdichte im Gebiet der oberen und mittleren Saale. Diss Univ Jena

Sauzay G, Courtois G (1970) Radioaktive Leitstoffe zur Messung von Schwebstoff- und Geschiebefracht. Int Symp Hydrometry Koblenz, Bes Mitt DGM 35:458–473

Savant SA, Reible DD, Thibodeaux LJ (1987) Convective transport within stable river sediments. Water Resour Res 23(9):1763–1768

Schad J (1913) Zur Entstehungsgeschichte des oberen Donautales von Tuttlingen bis Scheer. Jahresber Mitt Oberrhein Geol Vereins, NF Bd 3, H 2, Jg 1913:11–21

Schäfer W (1913) Die Flußdichte zwischen Teutoburger Wald und Wiehengebirge. Z Gewk, Dresden, 11(2):81–125

Schaefer I (1966) Der Talknoten von Donau und Lech. Mitt Geogr Ges München Bd 51, pp 59–111

Schaffernak F (1935) Hydrographie. Springer, Wien, 438 pp

Schaffernak F (1950) Grundriß der Flußmorphologie und des Flußbaues. Springer, Wien, 115 pp

Scharf W (1965) Petrographie der Gewässersohle. Ber Wasserwirtschaftl Rahmenplanung, Graz, 2:22–32

Scheidegger AE (1961) Theoretical geomorphology. Springer, Berlin Göttingen Heidelberg, 333 pp

Scheidegger AE (1967) A thermodynamic analogy for meander systems. Water Resour Res, Washington 3(4):1041–1046

Scheidegger AE (1973) Hydrogeomorphology (review). J Hydrol 20(3):193–215

Scheidegger AE, Langbein WB (1966) Steady state in the stochastic theory of longitudinal river profile development. Bull Int Assoc Sci Hydrol, Louvain, 11(3):56–61

Scheu E (1909) Zur Morphologie der Schwäbisch-Fränkischen Stufenlandschaft. Forsch Dtsch Landes-Volksk, Stuttgart, 18(4):361–404

Scheuerlein H (1984) Die Wasserentnahme aus geschiebeführenden Flüssen. W Ernst & Sohn, Berlin, 105 pp

Scheurmann K (1973) Die Pupplinger und Ascholdinger Au in flußmorphologischer Sicht. Wasser Abwasser (Bau-intern) 7:207–213

Scheurmann K (1977) Der Wasserkreislauf im Weltbild eines Gelehrten der Barockzeit. DGM, Koblenz, 21(2):21–27

Schlatte H (1979) Anhebung der Stromerzeugung durch Geschiebeprognosen auf Grund akustischer Geschiebemessungen. X. Konf d Donauländer über hydrologische Vorhersagen, Wien Abschn 33:1–14, Wien

Schmidt CW (1924) Der Fluß. Eine Morphologie fließender Gewässer. Dtsch Naturwiss Ges, Th Thomas-Verl, Leipzig, 77 pp

Schmitthenner H (1920) Die Entstehung der Stufenlandschaft. Geogr Z 26:207–229

Schmutterer J (1961) Geschiebe- und Schwebstofführung der österreichischen Donau. Wasser Abwasser, Wien, 1961:61–70

Schöberl F (1981) Abpflasterungs- und Selbststabilisierungsvermögen erodierender Gerinne. Österr Wasserwirtschaft 33(7/8):180–186

Schoklitsch A (1926) Geschiebebewegung in Flüssen und an Stauwerken. Springer, Wien, 108 pp

Schoklitsch A (1935) Stauraumverlandung und Kolkabwehr. Springer, Wien, 178 pp

Schoklitsch A (1938) Geschiebe- und Schwebforschung. Deutsche Wasserwirtschaft, Stuttgart, 33(8):188–192

Schoklitsch A (1942) Die Schweb- und Geschiebefracht italienischer Flüsse. Wasserkraft Wasserwirtschaft, München Berlin, 37(6):135–139

Schoklitsch A (1949) Berechnung der Geschiebefracht. Wasser- Energiewirtschaft 41:1–4

Schoklitsch A (1962) Handbuch des Wasserbaues. 2 Bde. Springer, Wien, 3. Aufl

Schröder RCM (1985) Vergleichbarkeit von Geschiebetransportformeln. Wasserwirtschaft, 75(5):217–221

Schröder W (1973) Bemessung des Regelungsgefälles alluvialer Sandbett-Bäche. Wasser Boden 25(11):351–356

Schumm SA (1957) Evolution of drainage systems and slopes in badlands at Perth Amboy, New Jersey. Geol Soc Am Bull 67:597–646

Schumm SA (1960) The shape of alluvial channels in relation to sediment type. US Geol Surv Prof Pap, 352-D

Schumm SA (1963) A tentative classification of alluvial river channels. US Geol Surv Circ 477:1–10

Schumm SA (1969) River metamorphosis. J Hydraul Div, Proc Am Soc Civ Eng 1:255–273

Schumm SA (ed) (1972) River morphology. Benchmark Pap Geol, Stroudsburg, Penn, 429 pp

Schumm SA (ed) (1977a) Drainage basin morphology. Benchmark Pap Geol Ser, Lond 41:352 pp

Schumm SA (1977b) The fluvial system. Wiley, NY, 338 pp

Schumm SA, Mosley MP, Weaver WE (1987) Experimental fluvial geomorphology. John Wiley, New York Chichester, 413 pp

Schwarzbach M (1974) Das Klima der Vorzeit. F Enke, Stuttgart, 3. Aufl, 380 S

Shahjahan M (1970) Factors controlling the geometry of fluvial meanders. IASH Bull 15(3): 13−24

Shalash S (1983) Degradation of the River Nile (International ab 1974). Water Power Dam Construction 35:37−43; 56−58

Shen HW (ed) (1979) Modelling of rivers. John Wiley & Sons, New York, 996 pp

Shen HW, Komura S (1968) Meandering tendencies in straight alluvial channels. Proc Hydrol Div, J Am Soc Civ Eng 94(6):997−1016

Shields A (1936) Anwendung der Ähnlichkeitsmechanik und der Turbulenzforschung auf die Geschiebebewegung. Mitt Preuß VA Wasser- Schiffbau, Berlin, 26:1−26

Shreve RL (1966) Statistical law of stream numbers. J Geol 74:17−37

Shreve RL (1967) Infinite topologically random channel networks. J Geol, Chicago, 75:178−186

Shreve RL (1974) Variation of main stream length with basin area in river networks. Water Resour Res 10:1167−1177

Shulits S (1941) Rational equation of river-bed profile. Am Geophys Union Trans 22:622−630

Sidki K (1980) Geologisch-Hydrogeologische Untersuchung zur Geometrie des Aquifers im Rhein-Neckar-Gebiet. gwf-Wasser/Abwasser 121(1):23−31

Sidle RC (1988) Bed load transport regime of a small forest stream. Water Resour Res 24(2):207−218

Simons BD, Şentürk F (1977) Sediment transport technology. Water Res Publ (ed) Ft Collins, 807 pp

Sindowski KH (1961) Mineralogische, petrographische und geochemische Untersuchungsmethoden. In: Bentz A (ed) Lehrbuch der Angewandten Geologie. 1. Bd Allg Methoden. F Enke, Stuttgart

Singer M (1913) Das Rechnen mit Geschiebemengen. Z Gewässerkde 4(4):239−272

Singh JB, Kumar S (1974) Mega- and giant ripples in the Ganga, Yamuna, and Son Rivers, Uttar Pradesh, India. Sediment Geol, Amst, 12:53−66

Sinnock S, Rao AR (1984) A heuristic method for measurement and characterisation of river meander wavelength. Water Resour Res 20(10):1443−1452

Sioli H (1965) Zur Morphologie des Flußbettes des Unteren Amazonas. Die Naturwissenschaften 52:104

Sirangelo B, Versace P (1984) Flood-induced bed changes in alluvial streams. Hydrol Sci J 29(4):389−398

Smart JS (1968a) Mean stream numbers and branching ratios for topologically, random channel networks. Bull Int Assoc Sci Hydrol 13(4):61−64

Smart JS (1968b) Statistical properties of stream lengths. Water Resour Res, 4:1001−1014

Smart JS (1969) Topological properties of channel networks. Bull Geol Soc Am 80:1757−1774

Smart JS (1978) The analysis of drainage network composition. Earth Surface Processes 3:129−170

Smart GM, Jäggi M (1983) Sedimenttransport in steilen Gerinnen. Mitt VA Wasserbau Hydrol Glaziologie, Zürich, 64:191 pp

Smith DI, Stopp P (1979) The river basin. In: Evans FC, Morgan MA (eds) Cambr Topics Geogr Ser. Cambridge Univ Press, Cambridge, 102 pp

Smith JD, McLean SR (1984) A model for flow in meandering streams. Water Resour Res 20(9):1301−1315

Smoltczyk KH (1955) Beitrag zur Ermittlung der Feingeschiebe-Mengenganglinie. Mitt Inst Wasserbau, TH Berlin, Berlin 43:62 pp

Snishenko BF, Kopaliani ZD (1984) Methods for estimating the impact of hydrotechnical structures on channel processes. Unpubl Pap

Sölch J (1949) Über die Schwemmkegel der Alpen. Ann Géogr, 31:369−383

Soni JP (1981) Unsteady sediment transport law and prediction of aggradation parameters. Water Resour Res 17(1):33−40

Sonntag R (1978) Schwebstoff-Führung und -Zusammensetzung in bayerischen Flüssen. Diss, TU München, 167 pp

Speight JG (1965) Meander spectra of the Angabunga River, Papua. J Hydrol 3(1):1−15

Spencer DW (1963) The interpretation of grain size distribution curves of clastic sediments. J Sed Petrol 33:180−190

Spengler R (1958) Der Stand der Diskussion über das Mäanderproblem. Geogr Ber, Berlin (Ost), 3(9):205–220

Sperling W (1957) Untersuchungen über Bettbeständigkeit und Feststoffbewegung im Flußbett. Die Wasserwirtschaft, Stuttgart, 47(12):301–305

Spöttle J (1902) Die schätzungsweise Bestimmung der Gesammtlänge der fließenden Gewässer im Königreich Bayern. II. Anhang zum Jahrbuch 1901 Kgl Bayer Hydrotech Bür, München, 10 pp

Stehr E (1975) Grenzschicht-theoretische Studie über die Gesetze der Strombank- und Riffelbildung. Hamburger Küstenforsch 34:188 pp

Stelczer K (1968) Der Geschiebeabschliff. Die Wasserwirtschaft, Stuttgart, 58(9):260–269

Stern R (1971) Kartierung von Wildbächen im Lesachtal (Kärnten). Carinthia II, Klagenfurt, 28:93–207

Sternberg H (1875) Untersuchungen über Längen- und Querprofil geschiebeführender Flüsse. Z Bauwesen 11/12:483–506

Stiny J (1910) Die Muren. Versuch einer Monographie mit besonderer Berücksichtigung der Verhältnisse in den Tiroler Alpen. Wagner, Innsbruck

Stiny J (1931) Die geologischen Grundlagen der Verbauung der Geschiebeherde in Gewässern. Springer, Wien, 120 pp

Strahler AN (1952) Hypsometric (area-altitude) analysis of erosional topography. Geol Soc Am Bull 63(11):1117–1142

Strahler AN (1953) Revisions of Horton's quantitative factors in erosional terrain. Am Geophys Union Trans (Abstr) 2:345

Strahler AN (1954) Quantitative geomorphology of erosional landscapes. CR 19th Intern Geol Cong, Algers, Sect 13, Part 3:341–354

Strahler AN (1957) Quantitative analysis of watershed geomorphology. Am Geophys Union Trans 38:913–920

Strahler AN (1968) Quantitative geomorphology. In: Fairbridge RW (ed) Encyclopedia of Geomorphology. Reinhold, NY, pp 898–911

Stratil-Sauer G (1950/51) Einige Vorbemerkungen zur Theorie der Erosion. Geol Bauwesen, Wien, 18(1):30–43

Stratil-Sauer G (1951) Stellungnahme zu einigen Auffassungen über das Flußlängsprofil. Sitzungsber Österr Ak Wiss, Math-nat-Kl, Abt I, Wien, 160:17–36

Strele G (1941) Bettbildung, Geschiebe- und Schwebstofführung. Wasserkraft Wasserwirtschaft, München, 36(8):204–205

Strickler A (1923) Beiträge zur Frage der Geschwindigkeitsformel und der Rauhigkeitszahlen für Ströme, Kanäle und geschlossene Leitungen. Mitt Eidg Amt Wasserwirtschaft 16

Sundborg A (1956) The River Klarälven: a study of fluvial processes. Geograf Ann, 38:127–316

Suerken J (1909) Die Flußdichte im östlichen Teile des Münsterschen Beckens. Diss TH Dresden

Sundborg A, Norrman J (1963) Göta älv, hydrologi och morfologi. Sveriges Geol Undersökn Arsbok, Ser Ca: Avhandl Uppsat 43:88 pp

Tarr WA (1924) Intrenched and incised meanders of some streams on the northern slope of the Ozark Plateau in Missouri. J Geol 32:583–600

Task Force for Bed Forms in Alluvial Channels (1966) Nomenclature for bed forms in alluvial channels. J Hydraul Div Am Soc Civ Eng 92(3):51–64

Thakur TR, Scheidegger AE (1970) Chain model of river meanders. J Hydrol 12(1):25–47

Thomas RB (1985) Estimating total suspended sediment yield with probability sampling. Water Resour Res 21(9):1381–1388

Thorarinsson S (1960) Der Jökulsa und Asbyrgi. Peterm Geogr Mitt, Gotha, 104(2/3):154–162

Thornbury WD (1969) Principles of geomorphology. Wiley NY, 594 pp

Thorne CP, Bathurst JB, Hey RD (1981) Direktmessung von Sekundärströmungen in natürlichen Mäandern. Wasserwirtschaft 71(10):283–288

Thorne CR, Bathurst JB, Hey RD (1987) Sediment transport in gravel-bed rivers. John Wiley & Sons, Chichester, 995 pp

Thornes J (1979) River channels. Aspects Geogr, Macmillan Education, Lond, 46 pp

Tietze KW (1961) Über die Erosion von unter Eis fließendem Wasser. Mainzer Geogr Stud, Panzer-Festschr, G Westermann Verl, Braunschweig Spec Iss, pp 125–142

Tietze KW (1975) Sedimentologische und geomorphologische Prozesse des Unteren Mississippi. Geol Rdsch 64(1):100–118

Tillmanns W (1984) Die Flußgeschichte der oberen Donau. Jh Geol Landesamt Baden-Württemberg 26:99–202

Timm J (1958) Anwendung der Schweizer Geschiebetriebformel. Die Wasserwirtschaft, Stuttgart, 48(15):415–418

Tinney ER (1962) The processes of channel degradation. J Geophys Res 67:4

Tippner M (1972) Beitrag zur Ermittlung von Gesetzmäßigkeiten der Geschiebebewegung im Oberrhein zwischen Freistett und Worms. DGM, Koblenz, 16(4):98–104

Tippner M (1973) Über den Umfang der Sohlenerosion in großen Gewässern. Dtsch Gewässerkundl Mitt 17(5):125–130

Tippner M (1976) Der Feststofftransport im Rhein. Beitr Rheinkde 27:32–38

Tippner M (1981) Meßtechnik für Naturmessungen. 2. DVWK-Fortbildungslehrg techn Hydraulik, Sedimenttransport in offenen Gerinnen, München-Neubiberg, 19 pp

Tison LJ (1968) Sedimentation und Erosion in Flußmündungen mit Gezeiten. Die Wasserwirtschaft 58(5):139–145; (6):183–188

Tödten H, Vetter B (1980) Untersuchungen zum Schwebstofftransport in natürlichen Gerinnen. Wasserwirtschaft 70(5):197–201

Trask PD (1932) Origin and environment of source sediments of petroleum. Gulf Publ, Houston (Texas), 323 pp

Troll C (1954) Über Alter und Bildung von Talmäandern. Erdkunde 8:286–302

Troll C (1957) Tiefenerosion, Seitenerosion und Akkumulation der Flüsse im fluvioglazialen und periglazialen Bereich. Peterm Geogr Mitt Erg, Gotha, 262:213–226

Trübswetter P (1930) Die Flußdichte des oberbayerischen Gebietes zwischen Inn und Salzach. Diss, Univ München, 81 pp

Türk B (1953) Untersuchungen über die Geschiebebewegung in Flüssen und Stauanlagen. Das elektroakustische Geschiebeabhörverfahren. Mitt Bl d BA f Wasserbau, Karslruhe, H 1:7–12

Tzschucke HP (1985) Verhinderung der Sohlenerosion unterhalb der Staustufe Iffezheim. Wasserbau-Mitt TH Darmstadt 24:57–69

Unbehauen W (1970) Die universelle logarithmische Geschwindigkeitsverteilung in natürlichen Gerinnen. Schriftenr Bayer Landesst Gewkde, München, 2:130 pp

US Department of Interior (1964) The nation's rivers. Government Printing Office, Washington DC

Vacher A (1909) Rivières à méandres encaissés. Ann Geogr 18:311–327

Vanoni VA (1941) Some experiments on the transportation of suspended load. Am Geophys Union Trans 22:608–620

Vanoni VA (ed) (1975) Sedimentation engineering. ASCE, NY, Manuals Rep Eng Pract 54:745 pp

Van Rinsum A (1950a) Die Schwebstofführung der bayerischen Gewässer. Beitr Gewässerkde, Festschr 50jähr Bestehen Bayer Landesst Gewässerkde, München, pp 103–110

Van Rinsum A (1950b) Der Abfluß in offenen natürlichen Wasserläufen. Mitt Geb Wasserbaues Baugrundf, Berlin, 7:80 pp

Van Rinsum A (1959) Querverbindungen in der Gewässerkunde. (Beziehungen zwischen Abfluß und Geschiebeführungen). DGM, Koblenz, 3(5):89–104

Vareschi V (1963) Die Gabelteilung des Orinoco. Peterm Geogr Mitt 107(4):241–248

Vetter M (1984) Die Anwendung der Gravitationstheorie zur Ermittlung der vertikalen Verteilung der Schwebstoffkonzentration. Mitt Inst Wasserw HSBw München, 13:171–203

Villinger E (1986) Untersuchungen zur Flußgeschichte von Aare-Donau/Alpenrhein und zur Entwicklung des Malm-Karsts in Südwestdeutschland. Jh geol Landesamt Baden-Württemberg, Freiburg/Brsg, 28:297–362

Vogt H (1963) Aspekte der Morphodynamik des mittleren Adour (SW-Frankreich). Peterm Geogr Mitt, Gotha, 107(1):1–13

Vollmers H, Pernecker L (1967) Beginn des Feststofftransports für feinkörnige Materialien in einer richtungskonstanten Strömung. Die Wasserwirtschaft, Stuttgart, 57(6):236–241

Von Baer KE (1860) Über ein allgemeines Gesetz in der Gestaltung der Flußtäler. Kaspische Studien VIII:1–6

Von Engelhardt W (1973) Die Bildung von Sedimenten und Sedimentgesteinen. Sediment Petrol, Stuttgart, 3:378 pp

Von Richthofen F (1901) Führer für Forschungsreisende. Verl v Robert Oppenheim, Berl 1886, 746 pp, Neudruck Hannover 1901, 734 pp

Von Riedl A (1806) Strom-Atlas von Baiern. Lentner'sche Buchhandlung, München

Von Schmidt-Kraepelin E (1973) "Peak Wilderness" — Wasserscheide der vier Ströme (Ceylon). Erdkundl Wissen (Beih Geogr Z), Wiesbaden, 33:352−397

Von Wissmann H (1951) Über seitliche Erosion. Colloq Geogr, Univ Bonn, 1:71 pp

Vossmerbäumer H (1976) Granulometrie quartärer äolischer Sande in Mitteleuropa. Z Geomorph 20(1):78−96

Vossmerbäumer H (1977) Allgemeine Geologie. E Schweizerbart, Stuttgart, 277 pp

Wagner G (1919) Die Landschaftsformen von Württembergisch Franken. Erdgesch Landeskdl Abh Franken Schwaben, Verl d Hohenloheschen Buchhandlung F Rau, Öhringen, 1:94 pp

Wagner G (1929) Junge Krustenbewegungen im Landschaftsbilde Süddeutschlands. Erdgesch Landeskdl Abh Franken Schwaben, Verl d Hohenloheschen Buchhandlung F Rau, Öhringen, 10: 302 pp

Wagner G (1955) Flußgeschichte, eine junge Wissenschaft. Aus der Heimat. Verl d Hohenloheschen Buchh F Rau, Öhringen, 63(7/8):134−148

Wagner G (1960) Einführung in die Erd- und Landschaftsgeschichte. Hohenlohesche Verlagsh F Rau, Öhringen, 3. Aufl, 694 pp

Walger E (1962) Die Korngrößenverteilung von Einzellagen sandiger Sedimente und ihre genetische Deutung. Geol Rdsch 54(2):494−507

Walger E (1965) Zur Darstellung von Korngrößenverteilungen. Geol Rdsch 54(2):976−1002

Walling DE (1988) Erosion and sediment yield research − some recent perspectives. J Hydrol 100(1/3):113−141

Wang SY, Shen HW, Ding LZ (eds) (1986) River sedimentation. ISRS Third Int Symp Proc Jackson Miss, The School of Engineering, The University of Mississippi, 3:1822 pp

Warntz W (1975) Stream ordering and contour mapping. J Hydrol 25(3−4):209−227

Weber H (1956) Gleichgewichtsgefälle und Erosionsterminante. Neues Jahrb Geol Paläontol 6:257−262

Weber H (1967) Die Oberflächenformen des festen Landes. BG Teubner, Leizpig, 367 pp

Wegener K (1925) Die theoretische Ablenkung der Flüsse durch die Erddrehung. Peterm Geogr Mitt, Gotha, 71:195−196

Wein N (1971) Auesande im Tal der Ems. Neues Arch Niedersachsen 20(4):336−347

Weiss FH (1986) Riverbed degradations downstream hydraulic structures, determination and possibilities for management. Third Int Symp River Sediment, Jackson Miss, 3:1124−1132

Werner C, Smart JS (1973) Some new methods of topologic classification of channel networks. Geogr Analysis 5:271−295

Werrity A (1972) The topology of stream networks. In: Chorley RJ (ed) Spatial analysis in geomorphology. Methuen, Lond, pp 167−196

Westrich B (1981) Verlandung von Flußstauhaltungen. Wasserwirtschaft 71(10):277−282

Westrich B, Kobus H (1983) Strahlerosion. Wasserwirtschaft 73(11):453−458

Whipple W Jr, Dilouie J (1981) Coping with increased stream erosion in urbanising areas. Water Resour Res 17(5):1561−1564

White WR (ed) (1982) Sedimentation problems in river basins. Stud Rep Hydrol, Project 5.3, UNESCO, Paris, 35:152 pp

White WR (ed) (1988) International conference on river regime Wallingford, UK. John Wiley & Sons, Chichester, 445 pp

Whitton BA (ed) (1975) River ecology − (studies in ecology). Blackwell, Oxford, vol 2, 725 pp

Wiberg PL, Smith JD (1987) Calculations of the critical shear stress for motion of uniform and heterogeneous sediments. Water Resour Res 23(8):1471−1480

Wilhelm F (1957) Flußmorphologische Untersuchungen in der Jachenau. Peterm Geogr Mitt, Gotha, Ergänzungsh 262:145−155

Wilhelmy H (1958) Umlaufseen und Dammuferseen tropischer Tieflandflüsse. Z Geomorph, NF, Berlin, 2:27−54

Wilhelmy H (1972/74) Geomorphologie in Stichworten. F Hirt, Kiel Bd 2, 223 S; Bd 3, 184 S; Bd 4, 375 pp

Williams GP (1986) River meanders and channel size. J Hydrol 88:147−164

Williams MAJ, Favre H (1980) The Sahara and the Nile. Maisonneuve Larose, Paris, 607 pp

Williams PF, Rust BR (1969) The sedimentology of a braided river. J Sediment Petrol 39:649−679

Wiltshire SE, Hewson AD (1983) A conversion factor for stream frequency derived from second series 1 : 25 000 scale maps. Inst Hydrol Rep, Wallingford, 84:9 pp

Winkel R (1942) Gefälleverstärkung eines Flusses durch Zufuhr von Schwemmstoffen. Die Bautechnik, Berlin, 20(3):31−32

Wittmann H (1942) Geschiebetrieb und Flußregelung. Dtsch Wasserwirtschaft, Stuttgart, 37(6):269−277, (7):333−339

Wittmann H (1955a) Flußbau. In: Schleicher F (ed) Taschenbuch für Bauingenieure. Springer, Berlin Göttingen Heidelberg, 2. Aufl, 2. Bd, pp 704−742

Wittmann H (1955b) Zur Morphogenese des Oberrheins. Die Wasserwirtschaft, Stuttgart, 45(5):121−131

Woeikof A (1885) Flüsse und Landseen als Produkte des Klimas. Auszugsweise Übers. des Kap. 8 aus dem Buch „Die Klimate des Erdballs", St Petersburg 1884 in Russian. Verlag Mittler u Sohn, Berlin, Z Ges Erdkde Berl, 20:92−110

Woldenberg MJ (1966) Horton's laws justified in terms of allometric growth and steady state in open systems. Geol Soc Am Bull 77:431−434

Woldenberg MJ (1969) Spatial order in fluvial systems: Horton's laws derived from mixed hexagonal hierarchies of drainage basin areas. Geol Soc Am Bull 80:97−112

Woldstedt P (1969) Quartär Handbuch der stratigraphischen Geologie. F Enke, Stuttgart, 2. Bd, 263 pp

Wolff W (1912) Die Flußdichte im Gebiet der Ahr, Erft und Roer. Diss Univ Bonn

Wolman MG (1971) The nation's rivers. Science, NY, 174:905−918

Woodruff J, Parizek EJ (1956) Influence of underlying rock structures on stream courses and valley profiles in the Georgia Piedmont. Assoc Am Geogr Ann 46:129−137

Wunderlich E (1929) Zur Entwicklung des Schmiech-Blautales. Z Ges Erdke 64:17−27

Wundt W (1941) Gefällskurve und Mäanderbildung als Folge des Prinzips des kleinsten Zwangs. Dtsch Wasserwirtschaft, Stuttgart, 36(3):115−120

Wundt W (1949) Die Flußmäander als Gleichgewichtsform der Erosion. Experientia 5:301−307

Wundt W (1952) Abtragung und Aufschüttung in den Alpen und dem Alpenvorland während der Jetztzeit und der Eiszeit. Erdkde, Bonn, 6(1):40−44

Wundt W (1953) Gewässerkunde. Springer, Berlin Göttingen Heidelberg, 320 pp

Wundt W (1962a) Zur Schwerstoff-Führung der Flüsse und Abtragung des Landes. Die Wasserwirtschaft 52(4):107−112

Wundt W (1962b) Aufriß und Grundriß der Flußläufe, vom physikalischen Standpunkt aus betrachtet. Z Geomorph, NF, Berlin, 6(2):198−217

Wundt W (1964) Die Erosion im Gebiete des Po mit anschließenden allgemeinen Folgerungen. Die Wasserwirtschaft 54(1):38−42

Yalin MS (1957a) Die theoretische Analyse der Mechanik der Geschiebebewegung. Mitt BA Wasserbau, Karlsruhe, 8:3−69

Yalin MS (1957b) Ermittlung des Querschnitts mit maximalem Geschiebetransportvermögen. Mitt BA Wasserbau, Karlsruhe, 9:40−57

Yalin MS (1960) Über die dynamische Ähnlichkeit der Geschiebebewegungen. Die Wasserwirtschaft, Stuttgart, 50(8):205−210, (9):244−248

Yalin MS (1972) Mechanics of sediment transport. Pergamon Press Oxford UK, 290 pp

Yang CT (1971) On river meanders. J Hydrol 13(3):231−253

Yang CT (1972) On river meanders, reply. J Hydrol 16(3):271

Yang CT (1974) Theories and relationships of river channel patterns, a discussion. J Hydrol 22(3−4):365−366

Yang CT (1979) Unit stream power equations for total load. J Hydrol 40(1−2):123−138

Yang CT, Song CCS, Woldenberg MJ (1981) Hydraulic geometry and minimum rate of energy dissipation. Water Resour Res 17(4):1014−1018

Yu B, Wolman MG (1987) Some dynamic aspects of river geometry. Water Resour Res 23(3):501−509

Zanke U (1977) Neuer Ansatz zur Berechnung des Transportbeginns von Sedimenten unter Strömungseinfluß. Mitt Franzius-Inst Wasserbau, TU Hannover, 46:157−178

Zanke U (1978) Zusammenhänge zwischen Strömung und Sedimenttransport. I. Berechnung des Sedimenttransportes − allgemeiner Fall −. Mitt Franzius-Ins TU Hannover, 47:214−345

Zanke U (1979a) Über die Abhängigkeit der Größe des turbulenten Diffusionsaustausches von suspendierten Sedimenten. Mitt Franzius-Ins TU Hannover, 49:245−255

Zanke U (1979 b) Über die Anwendbarkeit der klassischen Suspensionsverteilungsgleichung über Transportkörpern. Mitt Franzius-Ins TU Hannover, 49:256–263

Zanke U (1979 c) Konzentrationsverteilung und Kornzusammensetzung der Suspensionsfracht in offenen Gerinnen. Mitt Franzius-Ins TU Hannover, 49:264–282

Zanke U (1982) Grundlagen der Sedimentbewegung. Springer, Berlin Heidelberg New York, 402 pp

Zanke U (1987) Sedimenttransportformeln für Bed-Load im Vergleich. Mitt Franzius-Inst f Wasserbau u Küsteningenieurwesen, H 64:324–411

Zanke U, Bode E (1980) Neue Erkenntnisse im Sedimenttransport – Ergebnisse aus der Arbeit d Teilprojektes B 5 im SFB 79. Mitt Franzius-Inst TU Hannover, 50:251–275

Zavoianu J (1975) A morphometric model for the surface areas of hydrographic basins. Rev Roum Geol, Geophys Geogr: Geogr, Bukarest, 19(2):199–210

Zeller J (1963) Einführung in den Sedimenttransport offener Gerinne. Schweiz Bauz, 81(34):597–602, (35):620–626; (36):629–634

Zeller J (1965) Die „Regime-Theorie", eine Methode zur Bemessung stabiler Flußgerinne. Schweiz Bauz 83(5):67–72; (6):87–93

Zeller J (1967 a) Flußmorphologische Studie zum Mäanderproblem. Geogr Helv 22(2):57–95

Zeller J (1967 b) Meandering channels in Switzerland. Symp River Morph Bern I:174–186

Zernitz ER (1932) Drainage patterns and their significance. J Geol 40(6):498–521

Zimmermann C (1983) Einfluß der Sekundärströmung auf Sohlausbildung und Sedimenttransportvorgänge in offenen Gerinnen. Wasserwirtschaft 73(11):447–452

Author Index

Subject Index